BIOLOG

Population and Community Biology Series

Principal Editor

Michael B. Usher
Chief Scientific Adviser and Director of Research and Advisory Services, Scottish Natural Heritage, UK

Editors

D.L. De Angelis
Senior Scientist, Environmental Sciences Division, Oak Ridge National Laboratory, USA

B.F.J. Manly
Director, Centre for Applications of Mathematics and Statistics, University of Otago, New Zealand

The study of both populations and communities is central to the science of ecology. This series of books explores many facets of population biology and the processes that determine the structure and dynamics of communities. Although individual authors are given freedom to develop their subjects in their own way, these books are scientifically rigorous and a quantitative approach to analysing population and community phenomena is often used.

Already published

1. **Population Dynamics of Infectious Diseases: Theory and Applications**
   R.M. Anderson (ed.) (1982) 368pp. Hb.
2. **Food Webs**
   S.L. Pimm (1982) 219pp. Hb/Pb.
3. **Predation**
   R.J. Taylor (1984) 166pp. Hb/Pb.
4. **The Statistics of Natural Selection**
   B.F.J. Manly (1985) 484pp. Hb/Pb.
5. **Multivariate Analysis of Ecological Communities**
   P. Digby and R. Kempton (1987) 206pp. Hb/Pb.
6. **Competition**
   P. Keddy (1989) 202pp. Hb/Pb.
7. **Stage-Structured Populations: Sampling, Analysis and Simulation**
   B.F.J. Manly (1990) 200pp. Hb.
8. **Habitat Structure: The Physical Arrangement of Objects in Space**
   S.S. Bell, E.D. McCoy and H.R. Mushinsky (1991) 452pp. Hb.
9. **Dynamics of Nutrient Cycling and Food Webs**
   D.L. DeAngelis (1992) 285pp. Pb.
10. **Analytical Population Dynamics**
    T. Royama (1992) 387pp. Hb.
11. **Plant Succession: Theory and Prediction**
    D.C. Glenn-Lewin, R.K. Peet and T.T. Veblen (1992) 361pp. Hb.
12. **Risk Assessment in Conservation Biology**
    M.A. Burgman, S. Ferson and R. Akcakaya (1993) 324pp. Hb.
13. **Rarity**
    K.J. Gaston (1994) 224pp. Hb/Pb.
14. **Fire and Plants**
    W.J. Bond & B.W. van Wilgen (1995) 272pp. Hb.
15. **Biological Invasions**
    M. Williamson (1996) c. 256pp. Pb.

BIOLOGICAL INVASIONS

Mark Williamson

Department of Biology, University of York, UK

Published by Chapman & Hall, 2–6 Boundary Row, London SE1 8HN

Chapman & Hall, 2–6 Boundary Row, London SE1 8HN, UK

Chapman & Hall GmbH, Pappelallee 3, 69469 Weinheim, Germany

Chapman & Hall USA, 115 Fifth Avenue, New York, NY 10003, USA

Chapman & Hall Japan, ITP-Japan, Kyowa Building, 3F, 2-2-1 Hirakawacho, Chiyoda-ku, Tokyo 102, Japan

Chapman & Hall Australia, 102 Dodds Street, South Melbourne, Victoria 3205, Australia

Chapman & Hall India, R. Seshadri, 32 Second Main Road, CIT East, Madras 600 035, India

First edition 1996

Typeset in 10/12 Times by WestKey Ltd, Falmouth, Cornwall

Printed in Great Britain by T.J. Press (Padstow) Ltd, Padstow, Cornwall

ISBN 0 412 31170 4 (HB) and 0 412 59190 1 (PB)

A catalogue record for this book is available from the British Library

Library of Congress Catalog Card Number: 96-84897

∞ Printed on permanent acid-free text paper, manufactured in accordance with ANSI/NISO Z39.48-1992 and ANSI/NISO Z39.48-1984 (Permanence of Paper).

For Hugh, Emma and Sophia

Contents

A colour plate section appears between pages 116 and 117

Note: CFPs refer to the conceptual framework points given in Table 1.1, p.3 and repeated, as relevant, at the chapter heads.

Preface and acknowledgements

In a sense this book is a descendent of my *Island Populations* (1981). That led to my giving a paper on 'Darwin and Islands' at the 1983 Darwin Centenary meeting of the Linnean Society. From that, in 1984, came an invitation to join the SCOPE programme on the ecology of biological invasions. (The SCOPE programme is described in the Appendix to Chapter 1, p. 26.) Research on invasions has been one of my major interests ever since.

The two books have other points in common. The structure of both reflects third-year undergraduate lectures I gave at the University of York. Both have a mixture of ecology, genetics and evolution, of theoretical population ecology and natural history. Like Jacob Weiner (1995) I believe that 'individual ecologists should attempt to be both theoreticians and empiricists'. I hope the book will be of interest to undergraduate and graduate students, and to academics and other professionals in ecology and conservation. Parts of the book discuss biological control and the release of genetically modified organisms, and may be of interest to those concerned with research and administration in those areas.

My first and deepest thanks go to those who read the book in draft. Alastair Fitter, Richard Law, Julian Taylor, Charlotte Williamson and Gordon Woodroffe read all the chapters, Tom Fritts Chapter 5 and Terry Crawford and Nancy Knowlton Chapter 6. All made most helpful comments and suggestions, though as usual the views and mistakes remain my responsibility.

In the SCOPE programme, I would like to thank Hal Mooney, who chaired it, Francesco di Castri, Jim Drake, Richard Groves, Fred Kruger and Marcel Rejmánek who were the other members of the Scientific Advisory Committee, and Gordon Conway, Alan Gray, Martin Holdgate and Hans Kornberg who were the members of the British committee.

As I was starting to write this book, there was a workshop on invasion biology at the University of California, Davis. That was in May 1994, and I would like to thank the organizers, Jim Carey, Peter Moyle, Marcel Rejmánek and Geerat Vermeij for setting it up and for much advice and information since then, and all the participants of the workshop for discussion but especially Ted Grosholz, Alan Hastings, Dick Mack and Dan Simberloff for providing texts and references as well. Papers from the workshop are being published in *Ecology* and *Biological Conservation*.

Many other people have also helped. They were Jim Carlton, Bryan Clarke,

John Currey, Calvin Dytham, Hugh Evans, Nick Fielding, Simon Hardy, Peter and Sylvia Hogarth, Jim Hoggett, Barry Hughes, John Lawton, Mark Lonsdale, Norman Maitland, Andrew McLellan, Denis Mollison, Geoff Oxford, John Prendergast, Michael Shaw, Frank van den Bosch, Geoff Waage and Don Wood, and I thank them all.

Those who supplied pictures and drawings are acknowledged individually in the captions. While every effort has been made to trace copyright holders and obtain permission, this has not been possible in all cases.

Mark Williamson
Dalby, February 1996

1 A framework for the study of invasions

1.1 INTRODUCTION

Each April the lawns of the campus at the University of York turn blue with the flowers of *Veronica filiformis*, the slender speedwell (Plate 1). It is a striking sight, and biologically remarkable too. The plants are a self-sterile clone, and no seed is set, and so the flowers have no function. The plant is an invader from the Caucasus (Hultén and Fries, 1986), and it spreads from fragments of stems. Lawn mowers efficiently cut the stems into small pieces which are spread by the machines and shoes. Some think the sight a pretty one; others who think lawns should be green and hard-wearing regard it as a pest (Williamson, 1994). Its creeping habit adapts it to lawns, though it will grow in the longer grass of orchards and such places too.

Veronica filiformis was originally introduced as a rock-garden plant, in about 1920, but rapidly condemned as being too aggressive, over-growing other plants (Salisbury, 1961). It was first recorded growing wild in England in 1927, and has spread all over the British Isles (Bangerter and Kent, 1957, 1962), and over much of western Europe, over an area stretching from France to Norway to Estonia to Romania to the Alps, as well as to a few places in both the west and east of North America (Hultén and Fries, 1986). It is a simple example of a prominent and successful invading plant.

Biological invasion happens when an organism, any sort of organism, arrives somewhere beyond its previous range. Nowadays, most invasions come from human actions, deliberate or accidental. But natural invasions happen too, from minor changes of range to major invasions across continents. An example of such a major natural invasion is the collared dove, *Streptopelia decaocto* (see Figure 4.1). Similarly, most invasions are into habitats much affected by human actions, particularly into markedly disturbed habitats, but again there are invasions into natural habitats (which are all disturbed by natural processes to some extent).

Invasions and introductions have long fascinated many biologists. Darwin (1860), among a dozen or so references to the topic, especially noted instances where there had been marked ecological effects, though he never used the word pest. He said (p. 65): 'cases could be given of introduced plants which have become common throughout whole islands in a period of less than 10 years. Several of the plants, such as the cardoon and a tall thistle, now most numerous over the wide plains of La Plata, clothing square leagues of surface almost to the exclusion of all other plants, have been introduced from Europe' and 'Let it be remembered how powerful the influence of a single introduced tree or mammal [species] has been shown to be'.

Charles Elton, the father of animal ecology in Britain, with his broadcasts and book (1958) drove much of the interest and understanding of invasions in our lifetime. In the last decade, there has been a major SCOPE program on the ecology of biological invasions (Drake *et al.*, 1989), which produced 15 sets of reports (see the Appendix to this chapter, which says what SCOPE is). Among recent reviews of some sets of invaders are Cronk and Fuller (1995) on plants, Lever (1994) on vertebrates, both world surveys concentrating on pest species, and Mills *et al.* (1993) on all established species in the (American) Great Lakes.

One reason for studying invasions is that many invasive species have become serious pests. An estimate gives the cumulative losses in the United States from (some only) harmful non-indigenous species as being almost $100 billion by 1991 (US Congress OTA, 1993). But most successful, i.e. established, invaders are not pests, and most invaders are not successful. One theme of this book is that the reasons for this are not well understood and that prediction is difficult. Because of the economic importance of invaders, it is important to appreciate how primitive a subject ecology still is, to understand how little we understand. Predictions of the outcome of new invasions can only, as yet, be weak and unreliable. But we can be optimistic. Ecology is advancing fast, and the variable patterns of invasions are being better described and interpreted. It is possible to start putting numbers to some parts of the invasion process; my hope is that a better quantitative understanding of all the different parts will follow before too long.

While much has been written on invasions, most of it is case histories, surveys of case histories and statistically weak generalizations. All those are to be found here, but my intention is to give an over-view of the invasion process, from arrival, through spread to eventual equilibrium and, on the way, to take a critical look at what generalizations are possible. In this chapter I take four cases in detail, and relate them to the conceptual framework in which I think invasions can be studied. The four are a virus, a bird, a mammal and a set of flowering plants. Between them they give examples of the main points I wish to make in my conceptual framework.

1.2 THE CONCEPTUAL FRAMEWORK

The heart of this book is Table 1.1, which sets out the conceptual framework. At this point, you may well wish only to glance at it, and skip the exposition of this section and return to it later. It is easier to understand with real examples in mind, but conversely I want to describe my examples in the context of the conceptual framework. Points in the framework are labelled as, for instance, CFP 6, for conceptual framework point 6. If you have your own examples in

Table 1.1 The conceptual framework.
CFP indicates conceptual framework point

A Arrival and establishment

CFP 0	Most arrivals at present are from human importations, but natural arrivals are also of interest
CFP 1	Most invasions fail, only a limited number of taxa succeed (tens rule)
CFP 2	Invasion (or propagule) pressure is an important variable. So invasions are often to accessible habitats by transportable species
CFP 3	All communities are invasible, perhaps some more than others
CFP 4	The *a priori* obvious is often irrelevant to invasion success. Among factors to consider: r (intrinsic rate of natural increase), abundance in native habitat, taxonomic isolation, climatic and habitat matching, vacant niche

B Spread

CFP 5	Spread can be at any speed in any direction, in analysed cases usually either as predicted by estimates of r (the intrinsic rate of natural increase) and D (the diffusion coefficient), or faster

C Equilibrium and effects

CFP 6	Most invaders have minor consequences (tens rule)
CFP 7	Major consequences have as • effects: depressed populations to individual extinctions to ecosystem restructuring • mechanisms: enemies (vertical food-chain processes), competition, amensalism, swamping (horizontal food-chain processes)
CFP 8	Genetic factors may determine whether a species can invade; genetic factors affect events at the initial invasion; evolution may occur after invasion

D Implications

CFP 9	Invasions are informative about the structure of communitie and the strength of interactions, and vice-versa
CFP 10	Invasion studies are relevant to considering the risks of introducing new species or genotypes, the release of genetically engineered organisms and the success and consequences of biological control

mind, and you want to see how they fit this conceptual framework, remember that the framework points are statistical. They indicate what, from our present knowledge, seems usually to happen. Individual invasions can do almost anything and can behave in many strange ways. For each chapter after this one, the CFPs discussed there are listed at the chapter head.

There are 11 points arranged in four sets. The first point is numbered zero, as it is more a statement than an intellectual concept, leaving 10 conceptual points. It is possible to distinguish many possible stages of invasions, but I have chosen to group them into three, the beginning, the middle and the end. These sets are labelled A, B and C in Table 1.1; set D, with CFPs 9 and 10, covers the implications of invasions for both pure and applied ecology.

The beginning of invasions, set A, includes the arrival of a species in a new locality and its success or failure in establishing itself there. By establish, I mean establish a viable population, one that would be permanent if conditions did not change. Along with the important background consideration of CFP 0, there are four conceptual points in set A. CFP 1, that most introductions fail, is essential for assessing all other aspects of invasions. CFPs 2 and 3, that invasion (or propagule) pressure is important, and that all communities are invasible, are important considerations in assessing many of the generalizations that have been put forward. CFP 4, that the *a priori* obvious is often irrelevant, encapsulates the difficulties of predicting the success of invaders. CFP 4 reminds us that ecosystems are complex, that biological processes can be marvellously variable between species, and that factors that look trivial to the naïve observer can be of overwhelming importance to the species.

The middle stage of invasions, set B, starts once a species has established itself in a new place. In practice that may be at the start of set A, establishment, or later, but the process of spread, set B, is distinct conceptually if not in time. Spread is the most predictable and modellable part of invasions at the moment, so it is covered by just one conceptual point, CFP 5, and the most mathematical chapter of the book (Chapter 4). Slow spread should allow assessment of an invader and, if necessary, effective control measures against it. In practice, this seldom happens, and most control measures are applied as the stage covered by set C.

A spreading invasive species will eventually reach a limit, as in Figure 1.2, and its population cease to increase. That stage can be called, rather loosely, equilibrium, and in set C are three conceptual points relating to the ecological and other effects then seen. CFP 6, that successful invaders usually have minor consequences, echoes CFP 1, that most introductions are unsuccessful, and both points need to be remembered in discussions of the ecology and policy consequences of invasions. When an invasion does have major consequences, CFP 7 reminds us that these can be of all sorts, though the two major mechanisms, listed in order of importance, are more limited. The genetical side of ecology, often neglected but sometimes over-emphasized, is covered by CFP 8.

Table 1.1 is a framework, not the complete structure of points discussed in the book. Though in a sense the whole book is a framework, it is not intended to be a comprehensive review of invasions, but I hope you can bolt on your own examples and interests.

1.3 FOUR EXAMPLES

The four examples are a sea-bird, the fulmar *Fulmarus glacialis*, a mammal and a virus that attacks it, rabbit *Oryctolagus cuniculus* and myxoma, and a set of flowering plants, the four species of *Impatiens* or balsam established in Britain (three introduced and one native). Between them, they cover all the points in the conceptual framework. In each case, I will first describe the history, biology and geography of invasion, and then I will relate those to the conceptual framework and possible explanations of the success of each invasion. In all four cases there have been successful invasions. The statistics of failed invasions comes in the next chapter.

1.3.1 The fulmar

Most successful recent invaders have been moved by human action to disturbed habitats, but there are many invaders that have not. The first example is one such. The fulmar *Fulmarus glacialis* (Plate 2) is an oceanic sea-bird which has spread south from Iceland in a remarkable way in the last 200 years (Figure 1.1).

The fulmar looks somewhat like a gull; indeed, its name means foul gull, from the Scandinavian, no doubt from its habit of spitting oil from its stomach when approached. It is not a gull but a petrel, a tube-nose, a member of the order Procellariiformes. All the birds in this order are pelagic, feeding on fish and plankton and coming to land only to breed; they include albatrosses, shearwaters and storm petrels.

Fulmars are Arctic birds, about 0.5 m long with a wing span of about 1 m, weighing around 750 g. They are perhaps the most northerly breeding species of bird, breeding on cliffs in the extreme north of both the Atlantic and Pacific. The Pacific birds breed from the Kuril islands (South of Kamchatka) through the Aleutians and islands in the Bering Sea and along Alaska. They are usually put in a separate subspecies, *Fulmarus glacialis rodgersii*. Some populations are suffering from predation by introduced arctic and red foxes (*Alopex lagopus* and *Vulpes vulpes*) (Litvinenko and Shibaev, 1991), but there may still be two to three million pairs in the subspecies (Lloyd, Tasker and Partridge, 1991).

In the Atlantic Arctic there are several large colonies, between around 10 000 and 100 000 pairs, on Baffin Island, north-west Greenland, Jan Mayen (east of Greenland, north of Iceland), Bear Island (Bjornøya, between Norway and Svalbad), Svalbad and Franz Josef Land. There are smaller colonies in north-east Greenland and Nova Zemblya and in the other regions. For instance, in north-west Greenland, there are eight colonies ranging from a little less than 10 000 up to about 70 000, and four colonies with 50 or fewer birds,

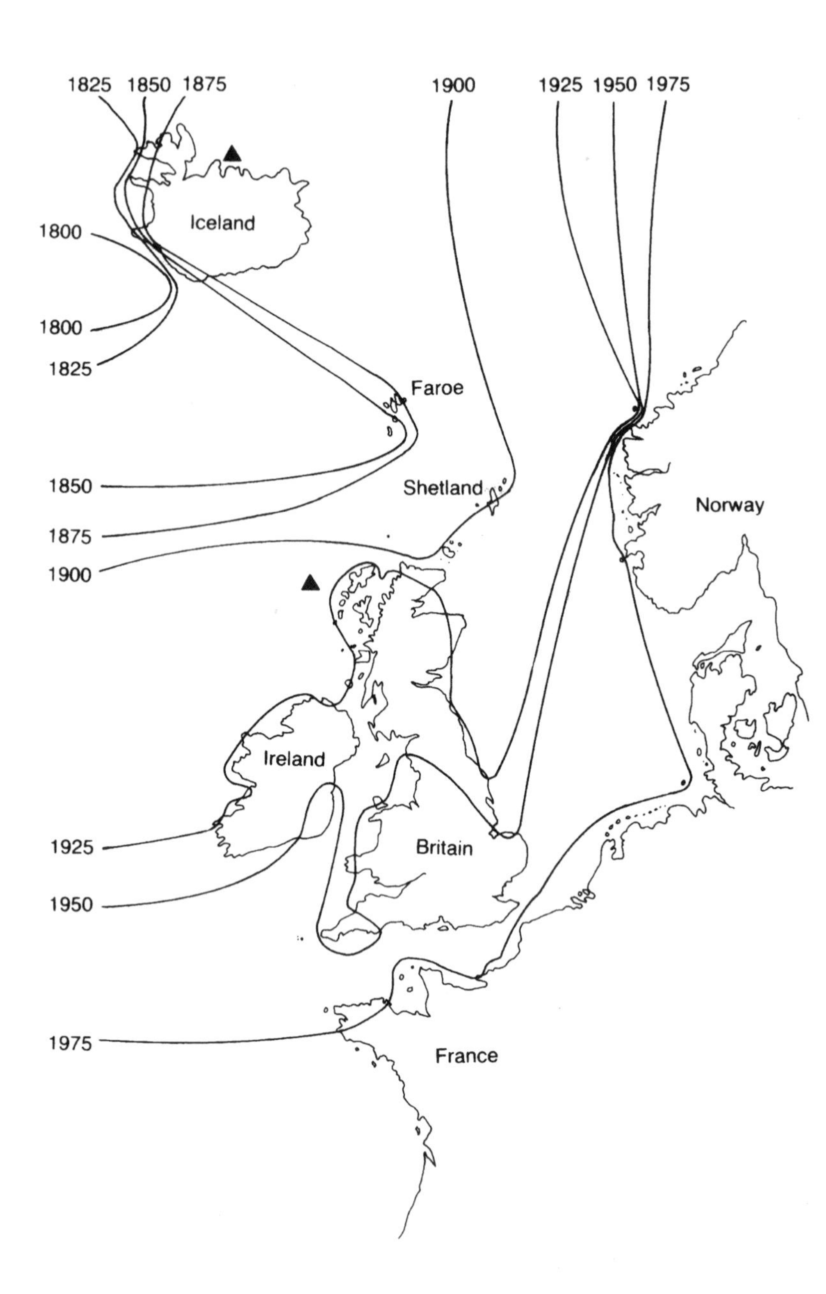

Figure 1.1 Map of the spread of the fulmar in the Atlantic. Contours at 25-y intervals. The triangles show the position of the two pre-spread colonies, Grimsey (no of Iceland) and St Kilda (west of Scotland). (Data from Cramp and Simmons, 19 Fisher, 1952, 1966; Toft, 1983; illustration by Mike Hill.)

none in between (Evans, 1985a). There are perhaps altogether between one-and-a-half to two million pairs in the Atlantic sector of the Arctic (Lloyd, Tasker and Partridge, 1991).

The only other species in the genus *Fulmarus* is the Antarctic fulmar, *F. glacialoides*, and the closest genera are also Antarctic. The fulmar is the only European bird to have all its closest relatives in the Antarctic (Voous, 1960). The Antarctic is not necessarily the place of origin for fulmars, as Miocene species (ca. 15 million years old) are known from California (Warheit, 1992). The Antarctic fulmar, like those from Britain, Iceland and north-west Greenland, is pearl grey above, white below, as are most of those on Jan Mayen and from the islands in the Bering Sea. However some populations are largely or entirely darker, mid to dark grey both above and below (Fisher, 1952). These include the Canadian arctic populations, those on Svalbad and Franz Josef Land, and all the other Pacific populations. The dimorphism is not entirely clear-cut, and its distribution is peculiar. The dark populations are the southernmost in the Pacific, the northernmost in the Atlantic and Arctic Oceans and the westernmost around Baffin Bay. But the distribution of the darker morph does limit where the spread shown in Figure 1.1 could have come from.

In the early 18th century there were only two sets of colonies of fulmars in the temperate North Atlantic, one around the island of Grimsey, about 400 ha, 40 km north of Iceland and exactly on the Arctic Circle, the other at St Kilda, a small archipelago of about 850 ha, over 50 km west of the Outer Hebrides of Scotland. Both of these are shown as triangles on Figure 1.1. The Westman islands (Vestmannaeyjar) off the south coast of Iceland have long been a major fishing centre. They are named from Irish slaves who, having killed their Viking masters in 871 AD, were themselves killed by other Vikings. Sometime between 1713 and 1757 fulmars started nesting on the Westman Islands (Fisher, 1952); the date is sometimes given as 1753, but that is merely the probable date of the first written record. There were fulmars on small islands further west at about the same time. Most were destroyed as fulmar sites by a volcanic eruption in 1783, but Eldey, where the last great auk *Pinguinus impennis* was killed in 1844, still has fulmars.

From these small beginnings off south-west Iceland, the fulmar has spread round Iceland, across to Faroe, then to Shetland. The first British breeding record, apart from St Kilda, is 1878 on Foula, the westernmost Shetland Island. The first known dates of breeding at colonies in Iceland, Faroe and Britain were meticulously recorded and mapped by Fisher (1952). He noted that the spread apparently by-passed St Kilda; it is as clear from his data that the spread by-passed Grimsey too. This astonishing spread has reached France, Germany (Helgoland) and Norway, as well as circumscribing the British Isles (Cramp and Simmons, 1977). The fulmar ranges far across the Atlantic and is commonly seen off Newfoundland (Brown, 1970) There are now small colonies in south-west Greenland (Evans, 1985a), Labrador and Newfoundland (Nettleship and Montgomerie, 1974; Montevecchi *et al.*,

Terms used in demography

Demography is the study of populations and their change. This box is intended as a reminder of the meaning of some terms in demography, and is not intended as a text on the subject.

The basic processes affecting the numbers of individuals in any populations are birth, death, immigration and emigration. Demography has to allow for changes in the rates of these over the life-cycle. It is usual to simplify at the start by ignoring immigration and emigration, and considering only females. With that, the two basic functions are:

survivorship function, l_a, the probability of survival from birth to age a,
maternity function, m_a, the expected number of female offspring per female aged a per unit time.

The **net reproductive rate**, R_0, is defined as the expected number of female offspring per female over a life time,

$$R_0 = \int_0^\infty m_a l_a da.$$

l_a and m_a will vary with environment and population density, but if they were fixed, not varying, the population would eventually increase at an exponential rate r, the **intrinsic rate of natural increase**. That is,

$$n_t = n_0 e^{rt}$$

where n_t is the population size at time t, n_0 is the initial population size, e is the base of natural logarithms and r is the intrinsic rate of natural increase, the only real and positive root of

$$1 = \int_0^\infty e^{-ra} l_a m_a da.$$

The derivation and meaning of this last equation are explained fully by Caswell (1989).

Although age and time are continuous, real calculations have to use discrete approximations, and many biological processes affecting population dynamics, such as breeding seasons or instars in insects, are more conveniently handled as discrete variables. So many find that the **Leslie matrix** model, using discrete variables instead of l_a and m_a easier to understand and to work with. The Leslie matrix is a square matrix of order s, where s is the number of stages, such as young, middle-aged and old, being considered. In the Leslie matrix the top row f_i gives the net fertilities allowing for mortality, the principal sub-diagonal p_i the survival probabilities. Here is a general Leslie matrix of order three:

$$\begin{bmatrix} f_1 & f_2 & f_3 \\ p_1 & 0 & 0 \\ 0 & p_2 & 0 \end{bmatrix}$$

In this form, $r = \ln \lambda_1$, where λ_1 is the dominant (the only real and positive) eigen-value of the Leslie matrix.

$$R_0 = f_1 + \Sigma_1^{s-1}(\Pi_1^{s-1} p_i f_{i+1})$$

or, for the matrix above,

$$R_0 = f_1 + p_1 f_2 + p_1 p_2 f_3.$$

For the logistic and other Lotka–Volterra equations, see the box in Chapter 7, p.181.

1978), and it may well be that these derive from the Icelandic–European spread rather than from Arctic colonies. Unfortunately there are as yet no genetic markers for these populations. Wynne-Edwards (1952) is sometimes quoted as showing a difference between St Kilda and other British birds; but, as he clearly stated, this difference was not statistically significant. He did have significant differences between three sets: Baffin Island birds, those on Spitzbergen, and those on Iceland, Faroe and Britain. That is, darker birds were slightly smaller; the Canadian birds, which are sometimes separated as the subspecies *F. g. minor* (Salomonsen, 1950; Godfrey, 1966; Howard and Moore, 1991), being the smallest.

The spread has been characterized by the founding of new colonies, and it would seem that the birds spread from these small colonies long before they grow to the size of Arctic colonies. The fulmar is a ocean-living bird, and the preferred nesting site is on exposed tall sea cliffs. As the population expands, these cliffs fill up. As Fisher (1952) says, 'ocean-facing cliffs are colonized first; then cliffs up-firth; then inland cliffs and lesser cliffs.' The direction and speed of the spread has been influenced by the distribution of cliffs, and perhaps by access to the open sea and its fishing grounds. But there are anomalies, most notably in Norway. A colony was founded at Rondøy, a well-known bird island near Ålesund, in about 1920 (Fisher, 1952), but there was no spread from there until some colonies were founded in Rogaland, near Stavanger, in 1968 (abandoned 1971), 1971 and 1973 (Toft, 1983). The position of these colonies is shown, somewhat inaccurately because of the contours, in Figure 1.1. Norway is well known both for its cliffs and its fishing industry, and the lack of spread there is surprising. I explain below why the fishing industry may be relevant.

The total population produced from this spread is known only very roughly, as the population in Iceland is given as 'ORD 6–7' (Evans, 1985b), i.e. between 100 000 and 10 million. There may be between 500 000 and 1 million pairs in Faroe, 1000 to 1500 in Norway, only a few tens in France, Germany, Newfoundland, Labrador and south-west Greenland. The 1985–1987 estimates for the British Isles are St Kilda 62 800, Ireland 31 300 and the rest of Britain 476 700 (Lloyd, Tasker and Partridge, 1991). The total number of pairs in the zone of spread is perhaps two-and-a-half million, comparable with the population of the Pacific subspecies and probably larger than the Atlantic Arctic populations.

The fulmar is a slow-breeding bird, laying at most one egg a year. From the long-term studies at Eynhallow on Orkney by Dunnet and his colleagues from Aberdeen, males start breeding at between 6 and 17 years, with a mode at 8, while females are even slower, between 7 and 19 years with a mode at 12 (Ollason and Dunnet, 1988). Fisher (1966) estimated the finite rate of increase, λ, to be about 5% per year, assuming first breeding at the age of 7. It certainly cannot be much higher than that and may be lower. So the intrinsic rate of increase, r, is less than 0.05 per year ($r = \ln \lambda$; if $\lambda = 1.05$, then $r = 0.048$). The rate of increase in numbers in parts of the British Isles in the early part of the spread was much higher than this, often around 20% (Fisher 1966), showing continued immigration from the north. The growth of the population over the whole British Isles has slowed from about 7% per year between 1939 and 1949 to about 4% from 1969 to 1986 (Lloyd, Tasker and Partridge, 1991). The intrinsic and finite rates, and other points in demography, are discussed in the box.

High arctic fulmars breed in large colonies. New colonies must have been established after the last glaciation, but the data suggest that when new colonies are formed they grow to a large size without budding off others. In contrast, in the spread south, small new colonies are formed both at the population front and behind it. Far more colonies were established in Iceland in the first half of the 20th century than in the 19th (Fisher, 1952). There is also some turnover in these new colonies. Between the two breeding atlas surveys of 1968–1972 and 1988–1991, 101 new 10-km squares were colonized, but breeding ceased in another 58 (Gibbons, Reid and Chapman, 1993).

Procellariiform birds feed on what can be caught at or near the surface of the open sea, plankton and nekton. Fulmars specialize on macro-plankton, objects a few centimetres across. Among such things now are the refuse, offal, from fishing and whaling. Fisher and Lockley independently came to the view that the spread of the fulmar resulted from modern fish trawling and the North Atlantic whaling that preceded it (Fisher, 1952, 1966; Fisher and Lockley, 1954). Gulls benefit from fishing near land, but it is primarily fulmars who benefit in the open sea.

There are two other theories for the spread of the fulmar (Lloyd, Tasker and Partridge, 1991). Salomonsen (1965) ascribed it to climatic warming, but it is difficult to see why that should benefit an arctic species. Wynne-Edwards (1962) thought there was either a genetic change, or a new fashion spread by imitation, favouring colonization, and that the spread was 'a wholly "natural" evolutionary phenomenon, basically unconnected with the activities of man'.

It certainly seems that new feeding and new breeding habits originated in south-west Iceland in the 18th century, and have been maintained throughout the spread. It seems plausible that this was a genetic change favoured by human activities. If so, the new genotype would have been attracted to the fishing industry, and so could have come from any of the pale fulmar populations: Grimsey, Jan Mayen, even West Greenland or St Kilda. Molecular genetic studies might help (Williamson, 1992). Some confirmation of this mixed theory comes from Furness and Todd (1984). St Kilda birds feed their chicks largely on macro-plankton, especially euphausiids (crustacea), but birds from Foula concentrate on fish (and offal).

In relation to the conceptual framework, fulmars show (CFP 0) that not all recent invasions are carried by man, though any fast modern invasion is likely at least to reflect anthropogenic habitat changes. Fulmars show the sort of surprising changes that result from all habitats being invasible, and that spread can be at any speed in any direction (CFPs 3 and 5). They have a slow rate of increase (CFP 4), and despite their huge numbers have had little measurable effect on other species (CFP 6). There may have been competition with other species for nesting sites on occasion, though it looks as though population interactions with other sea birds are minimal (CFP 9). The spread may possibly have involved a genetic change (CFP 8).

1.3.2 Rabbits and the myxoma virus

These next two examples, two and three, will be considered together. Both are well known but illustrate between them all the points in the conceptual framework.

Rabbits, *Oryctolagus cuniculus*, are a well-known pest, often living in dense colonies, grazing the vegetation closely and selectively, and so doing much damage. They are the only one of between 40 and 50 species of Leporidae, the hares and rabbits (Corbet and Hill, 1991), to be a widespread serious pest. The high density of some rabbit populations is a major cause of this. Indeed, some interesting leporids from the mountains of Mexico, from Nepal and Assam, from the Ryuku islands south of Japan, and from Sumatra, are endangered species.

The pathogen of myxomatosis is the myxoma virus, a poxvirus. This family of viruses is only known, as yet, from birds, mammals and insects. Poxviruses include fowlpox, smallpox and cowpox (though chickenpox is a herpes virus and so neither a poxvirus nor connected with chickens). Poxviruses are large for viruses, typically 200 nm across, are enveloped and contain double-stranded DNA (Fauquet, 1994). Myxomatosis was originally, and indeed still is, a disease of South American and Californian cottontail rabbits of the genus *Sylvilagus*. The strain that has been spread around the world to control rabbits comes from *S. brasiliensis*; there is a related strain in the Californian *S. bachmani*. In both these species the myxoma virus produces localized skin tumours around the point where it is inoculated by mosquito bites.

The history of myxomatosis shows the effect of changing host, and the speed at which a disease can spread. It shows some of the strengths and limitations of biological control. Both the virus and its host have evolved since they were first brought together. All these points are important in understanding the impact of invasions in general, and of genetically engineered organism in particular, a point I will return to in Chapter 6.

With a biological control agent it is important to know the host range There are about 12 species of *Sylvilagus*, cottontails or American rabbits native in North and South America (Corbet and Hill 1991); myxoma has bee tested in five of these and is always benign. Hares of the genus *Lepus* ar generally immune, so that their populations are not affected. Occasionally individuals of both species of hare found in Europe, the brown har *L. europaeus* and the mountain hare *L. timidus*, can catch the disease, usuall mildly but very rarely quite severely (Fenner and Ratcliffe 1965). But only i rabbits is myxomatosis a serious, often fatal, disease.

This change in pathogenicity is remarkable, important and as yet nc understood. Eventually, perhaps the molecular basis of pathogenicity will b known sufficiently well to predict what will happen to a pathogen introduce into a new species. For the moment, we must proceed by trial and error, an remain aware that a pathogen can become much more virulent in a new hos In general, pathogens will not affect, or will only affect mildly, new hosts t which they are introduced. The rare cases where the pathogen is more viruleı leads to some of the most spectacular invasions, as with myxoma, or wit chestnut blight (section 5.2.2). This epitomizes a major difficulty in predictin what will happen with new invasions. Although the probability of somethir unexpected happening is small, the consequences from such a happening ca be very large indeed. Rare, unexpected, events can be much more dramat than the predictable more normal ones. Nevertheless, there have been intr ductions which, at least in hindsight, might have been predicted to ha adverse effects. One of these is the introduction of wild rabbits into Austral (Figure 1.2).

Rabbits are natives of the Iberian peninsular. From the earliest times m has introduced rabbits to other areas, such as the Balearic Islands and Corsi (Cheylan, 1991; Flux, 1994). In mainland Spain, rabbits have never been pest. Compared with rabbits in other parts of the world, they are small mature earlier and have low fecundity (Jaksic and Fuentes, 1991). Rabbits a valued as stock that can use poor grazing land and they are used for both me and fur. The status of rabbits in much of Europe is that of game rather th vermin (Rogers, Arthur and Soriguer, 1994). In Western Europe, many intr ductions north of Spain came possibly not from there but from stocks dom ticated in French monasteries in the dark ages (Flux, 1994).

Rabbits were introduced into England in the 12th century, probably fro France. Rackham (1986) recounts how they were originally delicate, had have burrows built for them, and were a luxury. He may overstate the ca There were records of damage in England as early as the 14th century, and that time they were being farmed systematically, for example on t Pembrokeshire Islands (Lockley, 1947). Over the centuries, rabbits beca apparently hardier and certainly more common. These changes were probal partly the result of improvements in agriculture, in particular the developm

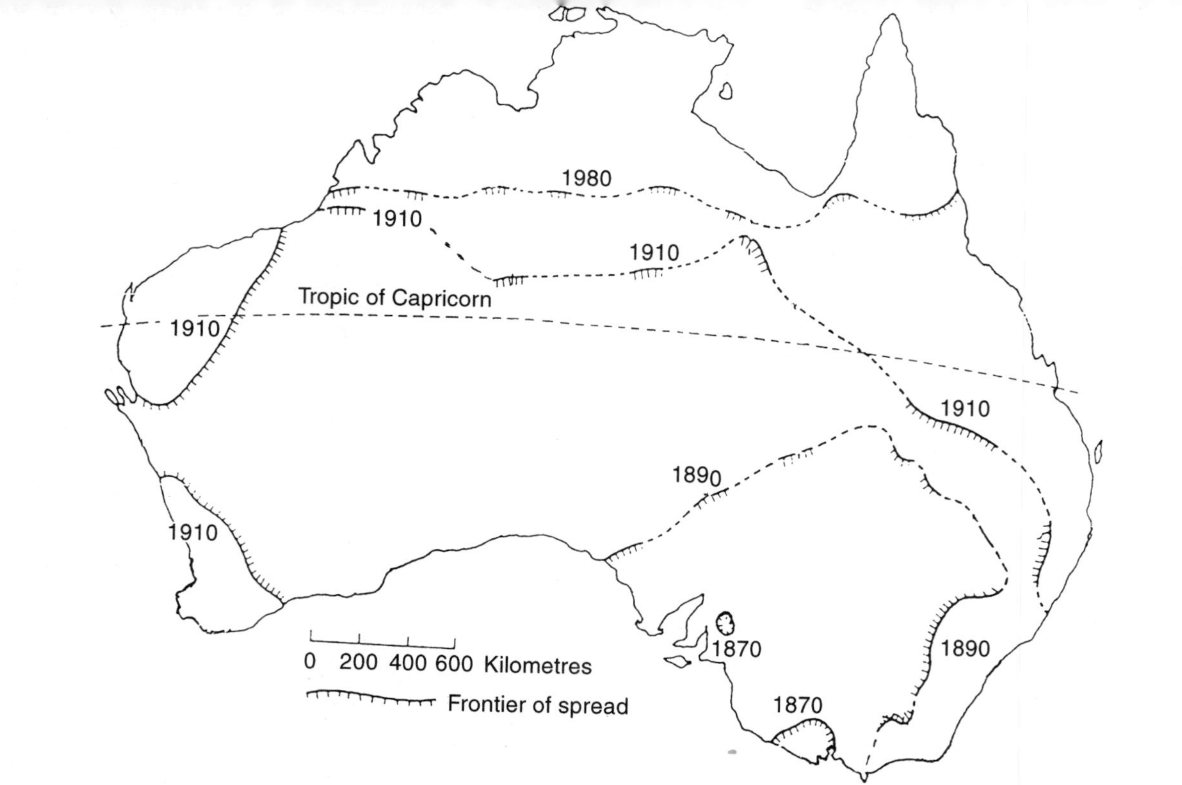

Figure 1.2 Simplified map of the spread of the rabbit in Australia. Contours for 20-year intervals for 1870, 1890 and 1910, and for the apparent limit of spread by 1980. (Modified from Stodart and Parer, 1988, with the permission of the CSIRO Division of Wildlife and Ecology.)

of winter crops. Rabbits were widely regarded as a pest from the mid-18th century, and were a staple food for the poor by the 19th century. Rackham (1986) is of the opinion that there was a definite, genetic, evolution of the rabbit in England, a proposition that is difficult to prove or disprove. These are all wild rabbits. Domesticated strains were also developed, larger, more placid and with various coloured furs.

Domesticated rabbits were introduced into Australia repeatedly in the early 19th century, the first five coming with the First Fleet in 1788. Wild-type, if not wild, rabbits were also introduced more than 30 times (Stodart and Parer, 1988). The main origin of the plague of rabbits was 24 (or fewer) imported on the clipper ship *Lightning* in 1859 and released at Barwon Park near Geelong, Victoria. These were mostly, though not entirely, genuinely wild (Fenner and Ratcliffe, 1965). The main spread started in Victoria and South Australia, as shown in Figure 1.2. The spread was largely natural, by reproduction and dispersal, and reached 130 km yr^{-1}, though it was helped by other releases too. It was fastest in low rainfall areas of 'shrub steppe, scrub savannah and low tree savannah' (Stodart and Parer, 1988). That is a far cry from the English landscape, and much drier even than Spain. In the wetter and more densely wooded mountain ranges the spread was slower. The 1980 limit of Figure 1.2 is, except in the eastern quarter, where the summer rainfall increases to 400 mm yr^{-1}. Less exactly, it is where tussock grassland, Acacia scrub and spinifex changes into savanna woodland (Blakers *et al.*, 1984; Lonsdale, 1994). The extent of agriculture in Queensland means that the boundary there is further north than would be expected from the climate and the native vegetation.

The main reason for importing rabbits to Australia was to shoot them as game. In South Australia rabbits were protected for 4 months of the year in the Games Act of 1864. The increase and spread, when it started, was rapid, and there was a South Australian Rabbit Destruction Act in 1875. Some farms were abandoned because of rabbits as early as 1881. Although some spread came from other points of introduction than Barwon, notably Kapunda in South Australia as shown in Figure 1.2, the major pattern is a single wave up to 1910. To the north, there was only a little spread from then to 1980. The lines shown in Figure 1.2 are somewhat misleading. Rabbits north of the Tropic of Capricorn are found as scattered colonies, becoming more isolated further north. It is unlikely there will be further spread, though climate change and land-use change could affect that. The spread at the east of the northern limit after 1910, through Queensland, came from expanding agriculture in the 1960s and 1970s (Myers *et al.*, 1994).

In north-western Europe, rabbits are found from Britain to Poland, from south Sweden to the south of France. In the Mediterranean, apart from the native Spain and Portugal, rabbits are found in northern parts of Morocco and Algeria, in scattered parts only of Italy, and on various islands as far east as Crete (Flux, 1994). The American eastern cottontail, *Sylvilagus floridanus* was introduced for sport after myxomatosis reduced the European rabbits, and is now widespread in France, Italy and northern Spain (Macdonald and Barrett, 1993). Rabbits have been introduced to Chile (Jaksic and Fuentes, 1991) (south of the area of the native *Sylvilagus brasiliensis*), to New Zealand, Tasmania and to over 800 small islands and island groups throughout the world (Flux, 1994).

Tasmania and New Zealand are both much more like England in their climates than most of Australia. Most 19th century introductions to these two places were of domestic rabbits and most, but not all, failed or failed to spread. In New Zealand, all rabbits seem now to be wild-type, but in Tasmania there are still domestic genes in at least some populations (Richardson, Rogers and Hewitt, 1980). Rabbits were serious pests in both by the 1870s. In New Zealand they were brought under control by an integrated programme starting in 1947. Important aspects were the establishment of local committees and decommercialization, so that there was local involvement and no one had an incentive to encourage rabbits. Control was by better farming and poisoning; shooting was encouraged but probably had little effect. Myxoma was never introduced (Gibb and Williams, 1994).

On small islands, rabbits have sometimes died out, sometimes been exterminated, and have quite often become a pest. The records are regrettably not good enough to allow analyses of frequencies, climatic correlates, or causes. An extreme example is the Hawaiian island of Laysan, a sub-tropical raised atoll. There was then a population explosion, turning the island into a desert, and exterminating two of three endemic land bird species. The rabbit population then starved, and all the few survivors were shot, so the vegetation has recovered (Williamson, 1981).

Rabbits show that pests are often introduced deliberately, and can be difficult to control. One approach, which works only in a minority of cases (see Table 2.6) is biological control, the introduction of predators parasites or pathogens to reduce the population. In general, acceptable biological control only comes from species-specific agents. As was noted above, myxoma is not specific to rabbits, but only in rabbits does it normally cause death. So, myxoma seems a possible good control for rabbits. The major point against it is that its symptoms are unpleasant to see (Plate 3), which is one reason why it has not been introduced into New Zealand.

Myxomatosis was first found in laboratory rabbits in Montevideo, Uruguay in 1896. The origin of the disease from *Sylvilagus brasiliensis* was only established by Aragao, in the 1940s. Earlier, in 1919, he had alerted the Australian government to the possibility of myxomatosis as a control agent for rabbits, but this came to nothing. In 1934, a Melbourne poliomyelitis specialist, Dr (later Dame) Jean Macnamara, urged the Australian government to try myxomatosis. By now, the problems from introduced species were all too well-known, and the import of the virus was not allowed. Nevertheless, tests by Sir Charles Martin at Cambridge in England were financed. These led to two field trials. The first was on the Pembrokeshire island of Skokholm, Wales, in 1936 and 1937 and failed. The second, a series of trials in south

Wales, in 1936 and 1937 and failed. The second, a series of trials in so
Australia, ending in 1943, was not a success, though there was some mi
temporary, spread of the disease.

Dame Jean Macnamara urged further trials, which were done in 195
the Murray Valley, in Victoria and New South Wales. The initial results
five sites, again indicated that the disease would not persist. Suddenly
December 1950, an epidemic started at just one of the sites. In 1951 it spr
widely in the Murray Valley, and over a large area along the boundary betw
New South Wales and Queensland. By 1953 it occurred, albeit sporadica
throughout the southern half of Australia, and also in Tasmania. The v
has been endemic (i.e. permanently established) in Australia ever since (Fen
1965, 1983; Fenner and Ratcliffe, 1965; Fenner and Ross, 1994).

The European epizootic started in two rabbits on the estate of Dr Del
at Maillebois in France, not far from Paris, in 1952. He used a strain that
been isolated in Brazil in 1949, and his trial was instantly successful. My
matosis spread throughout France in 1953 and reached Britain that year.
the end of 1955 it was throughout England and Wales. As with the sprea
the rabbit in Australia, the spread of myxomatosis was often speeded by m
Farmers anxious to rid their land of rabbits introduced the disease.

The critical factor in both the success and failure was the transmissio
the disease. The disease is spread by a variety of biting insects. The tr
mission is purely mechanical, there is no reproduction in the insect.
Australia transmission is dependent on an abundance of mosquitoes, *A
pheles*, *Culex* and others. The native flea *Echidnophaga myrmecobii* is kn
as the stick-fast flea and is not an effective vector. In Britain, transmissio
primarily by the rabbit flea, *Spilopsyllus cunciculi*. In France, both mosqui
and fleas can be important. The Skokholm trial failed because there wer
rabbit fleas there. The first Australian trials fizzled out from lack of vect
Although the disease is endemic throughout Australia, it is not clear ho
persists through droughts, when vectors disappear over large ar
Spilopsyllus was introduced to Australia in 1968, with little effect (Pa
Conolly and Sobey, 1981), though it is established as a vector in areas v
more than 200 mm rainfall a year (Myers *et al.*, 1994). Arid adapted Spa
fleas are now being tried (D. Wood, personal communication).

Initially, over 99% of the rabbits died; with that death rate rabbits wo
be expected to be very rare, if not extinct. That did not happen, though rab
are mostly much less common than they used to be. What happened?

From the initial state, where over 99% of the rabbits died from my
matosis, there was evolution, in both Australia and Britain, to an equilibr
where 30–50% survived. The details were shown by some brilliant work
Fenner, summarized in Table 1.2. Using standard laboratory rabbits to as
the virus, and a standard virus strain to assay rabbits, Fenner showed that
individual myxoma strains could be characterized by the mortality produ
or the mean survival time of rabbits. He classified strains from I, aln
invariably fatal, to V, mild. The original strains of myxoma released v

Table 1.2 Degrees of virulence* in myxomatosis and their frequency in the field

	Grade					
	I	*II*	*IIIA*	*IIIB*	*IV*	*V*
Mortality (%)	>99	95–99	90–95	70–90	50–70	<50
Mean survival (days)	<13	13–16	17–22	23–28	29–50	–
Recovery rate (per day)	0.0001	0.0022	0.0042	0.0100	0.0169	0.0301
Britain 1962 (%)	4.1	17.6	38.8	24.8	14.4	0.5
Australia 1963/1964 (%)	0	0.3	26.0	34.0	31.3	8.3

*Virulence = (mean survival)$^{-1}$; the grades are defined by the mean survival. From Fenner and Ratcliffe (1965) and Anderson and May (1982).

grade I, the other grades appeared in the field, by evolution, quite quickly. An equilibrium was reached with grades IIIA and IIIB predominating, but with all grades present, as can be seen in Table 1.2, which gives the definitions of the strains. That these proportions represent an equilibrium is shown by latter results given by Fenner and Ross (1994), though they do not distinguish there between the two types of strain III. There was similar evolution in France. There, by 1968, more than half of the myxoma strains were IIIB and IV, and that proportion was maintained until 1978. Since then the average virulence has actually increased (Rogers, Arthur and Soriguer, 1994).

At the same time there was evolution of the rabbits. The results for the reaction of rabbits at Lake Urana to a standard IIIA myxoma are shown in Figure 1.3. In 6 years, the population changed from one in which only a few percent survived, to one in which almost half survived, and about one-third had only mild symptoms. In Britain, rabbits are now about one-third to one-half of their number before myxomatosis. In Scotland, 16.7% of farms now have severe infestations compared with 53.7% before myxoma was introduced (Kolb, 1994). A little is known of the genetics of resistance, and again it seems as though an equilibrium has been reached at an intermediate state (Fenner and Ross, 1994).

This evolution to an intermediate state was unexpected. Conventional wisdom was that there would be co-evolution to a mild state. In fact, no particular equilibrium can be predicted *a priori.* As Anderson and May (1982) say 'the complicated interplay between virulence and transmissibility of parasites leaves room for many coevolutionary paths to be followed, with many endpoints.'

The course of an epidemic is determined largely by the basic reproductive rate R_0 (see the box) of the pathogen (or parasite). For a pathogen this is measured as the average number of secondary cases of infection generated by one primary case in a susceptible host population. R_0 is not a constant, but a function of the transmission rate, the virulence, the recovery rate, the disease-free mortality rate and the population size of the host. Using Fenner's data, and particularly the relationship between virulence and recovery rate in Table 1.2, Anderson and May (1982) calculated an intermediate grade of myxoma

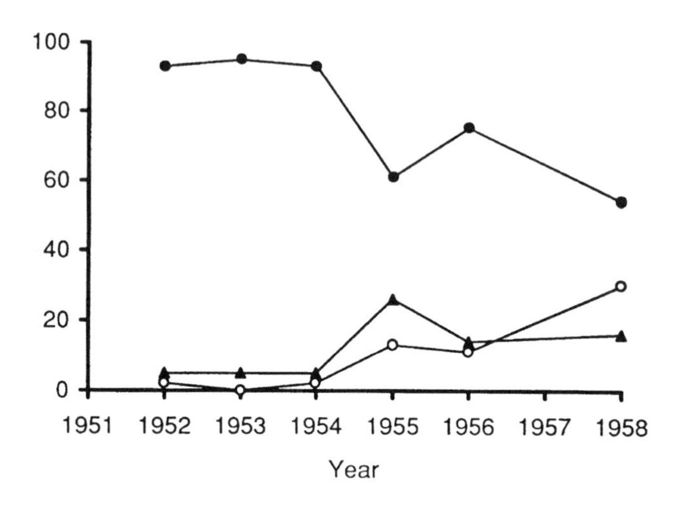

Figure 1.3 Change in the severity of myxomatosis in rabbits from Lake Urana, NSW, Australia challenged with a standard IIIA strain of myxoma. Data from Fenner and Ratcliffe (1965) p. 246. 1952 is before myxoma reached the area. ●, severe and fatal; Δ, moderate; ○, mild. The ordinate is the percentage of rabbits falling into the three

would have the highest R_0, giving at least a first-order explanation of the evolutionary change. Roughly, if the strain is too virulent, the hosts die so fast that transmission is limited. Conversely, if the strain is mild, giving a quick recovery, transmission is again reduced. Strains with intermediate virulence are transmitted best.

Fenner's work was done in the 1960s, so nothing is known as yet of the molecular changes involved in the evolution of either myxoma or rabbit. Introduced rabbit populations have evolved in other ways. In Tasmania, there is a balanced genetic polymorphism with a high frequency for the well-known gene for black coats (Barber, 1954). This variation follows Gloger's rule, cool, damp climates favouring darker forms. Similarly, on Skokholm, rabbits are smaller, run more slowly and there is a higher frequency of blacks than on mainland Wales (Lockley, 1947).

How do rabbit and myxoma relate to the conceptual framework (Table 1.1)? Both are human introductions (CFP 0) and in both cases many of the initial introductions failed (CFP 1). The critical factors may have included, for rabbits, stock with genes appropriate for living free and, for myxomatosis, timing in relation to vectors, but in both cases there would not have been successful invaders if those introducing them had not been persistent (CFP 2).

Why rabbits are invasible by myxoma, and why rabbits succeed in a range of habitats far beyond those found in Spain, are currently beyond explanation. These empirical facts relate to the proposition that all communities are invasible (CFP 3). The range of possible species, and the physiological limits of pecies, are both so large that it should always be possible to find a species apable of invading any specified habitat. Nor is it clear that there is any articular reason (CFP 4) for rabbits being a successful invader and a pest outside Spain) other than that they have been carried to ecosystems that suited hem and have reached high densities, which is almost tautologous.

Rabbits spread very fast across the centre of Australia, slowly in the nountains and not at all into the wetter tropical areas of the North, demontrating (CFP 5) the variability of observed spreads. CFP 6, that most invaders ave minor consequences, is shown, a little obscurely, in various ways. Rabbits re the only pest in the lagomorphs, myxoma is an exceptionally lethal disease. ome island rabbit populations had little impact, indeed died out naturally Flux and Fullagar, 1992), and some myxoma/rabbit strain combinations have benign pathology. But in general, this pair of species show CFP 7, major ffects through vertical food-chain processes, the rabbit on the vegetation, nyxoma on the rabbit. Both show the effects of selection following an invasion CFP 8, Table 1.2, Figure 1.3).

The last two CFPs (9 and 10) need little comment. These are two exceptional nvaders, with exceptionally strong interactions, but both show how difficult t is to foresee clearly the effects of an invasion.

1.3.3 *Impatiens*

he fourth and final introductory example is a set of species of *Impatiens*, nnual herbs known variously as touch-me-not, balsam, busy lizzie and ewelweed, among other names. *Impatiens* is a large genus of perhaps 850 pecies (Mabberley, 1987), comprising all but one of the species in the alsaminaceae. Most of the species are tropical and sub-tropical African and sian, and some of the non-temperate ones are perennials and semi-shrubs. ouch-me-not is a name because the ripe seed capsule bursts explosively when ouched, throwing the seeds a metre or two from the plant. Many species have een grown in botanic gardens and perhaps a dozen are available commerially in Europe. Only three have invaded semi-natural and natural habitats n Europe. These are *I. glandulifera* or Himalayan Balsam (Plate 4), *. parviflora* or Small Balsam and *I. capensis* or Orange Balsam. The native pecies is *I. noli-tangere* touch-me-not balsam (Plate 4). All four are shown in igure 1.4.

The most widespread now of these three invaders is *I. glandulifera*, which s curious as it has the smallest native range of the three. As we will see later section 3.4.3), there is in general a definite, if weak, three-way relationship etween native range size, abundance and success at invasion. Its natural range Figure 1.5) is in the western Himalaya, perhaps 800 km long by 50 km wide, etween Kashmir and Garwhal (in the Indian state of Uttar Pradash) (Gupta, 989; Polunin and Stainton, 1984). It is found in 'Shrubberies, bushy places; ommon on grazing grounds; often growing gregariously' (Polunin and tainton, 1984), mostly at altitudes between 2000 and 2500 metres.

In Britain, it is found on 'the banks of rivers and canals, damp places a waste ground' (Stace, 1991) including damp woodland. It can be a pest woodland, forming mono-specific stands in the herb layer (Perrins, Fitter a Williamson, 1993), but can be an attractive and welcome plant with its la purple flowers along otherwise dreary canal banks. Another name for it

Figure 1.4 Four species of *Impatiens*. Top left, *I. parviflora* with yellow flowers; right, *I. glandulifera*, flowers pinkish-purple (see Plate 4); bottom left, *I. capensis*, ora flowers with brownish blotches; bottom right, *I. noli-tangere,* yellow flowers (see P 4). Note the small closed cleistogamous flowers at the top of the lower two. (Illustra by Mike Hill.)

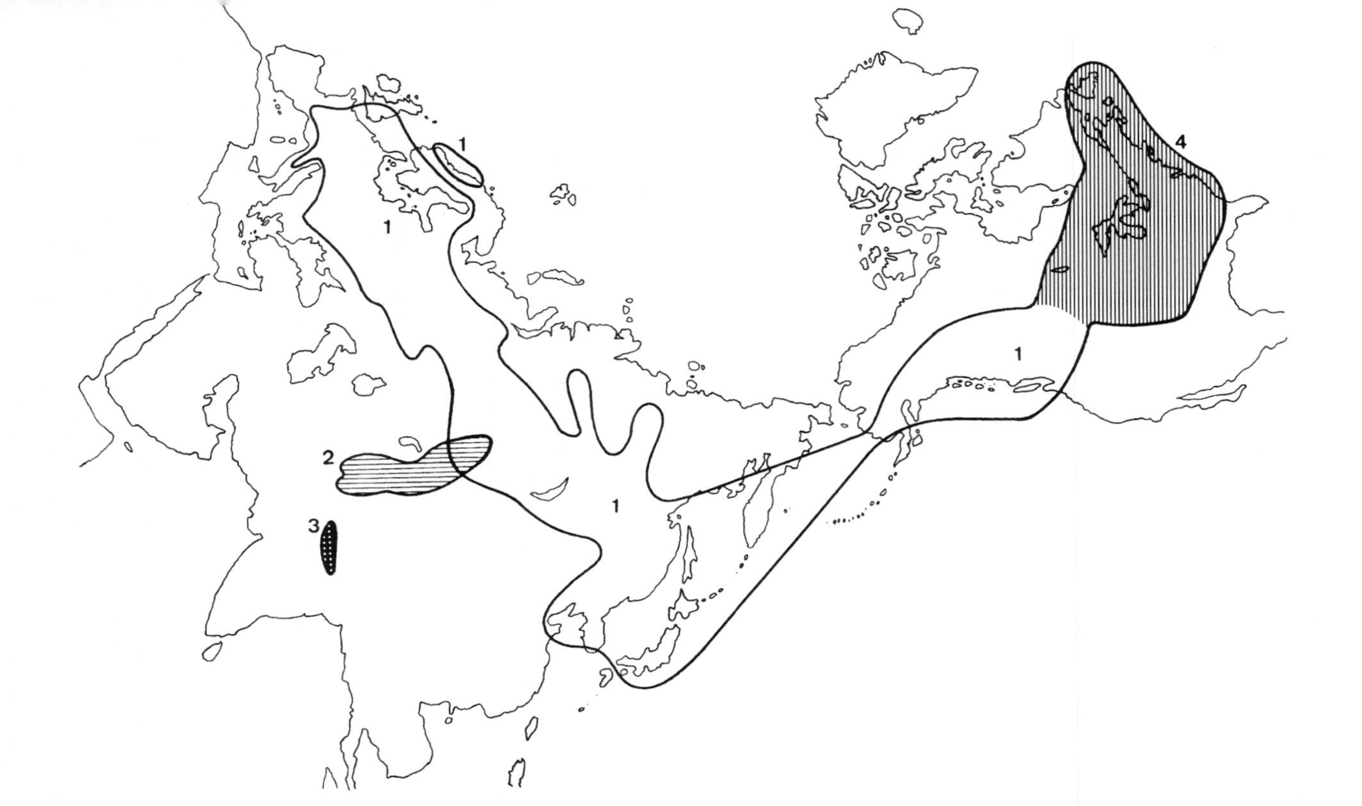

Figure 1.5 Map of the native distribution of four species of *Impatiens*. Europe to the left, North America to the right. Many people find this map easier to grasp if they rotate the book. Species 1, *noli-tangere*; 2, *parviflora*; 3, *glandulifera*; 4, *capensis*. (Illustration by Mike Hill.)

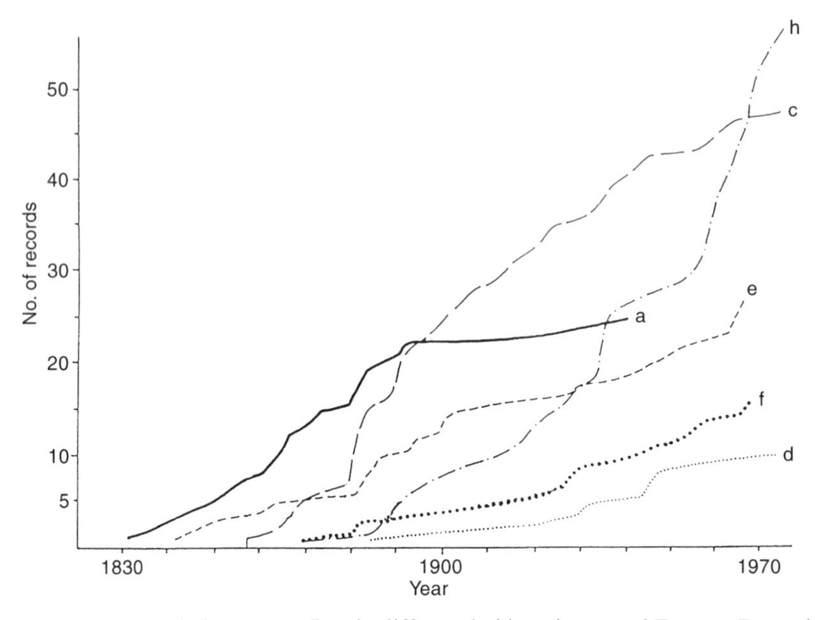

Figure 1.6 Records for *I. parviflora* in different habitats in central Europe. Records of, and with permission from, Dr L. Trepl. Simplified from Kornaś (1990). a, botanical gardens; c, parks and cemeteries; d, industrial plants and timber yards; e, gardens and hedgerows; f, ruderal places; h, natural forests.

Policeman's Helmet, from the flowers. It is now the tallest British annual plant (Beerling and Perrins, 1993); it grows to 2 m high. It has also invaded Ireland, continental Europe from France to Central Europe to southern Scandinavia (Fitter, 1978), and parts of Ontario, Quebec, New England and the maritime provinces in North America.

I. parviflora, with small yellow flowers, has, like *I. glandulifera*, invaded central Europe and southern Scandinavia including parts of Finland, but it is not found in Ireland, and only at a very few localities in Quebec and Prince Edward Island in North America. Its usual height is about 60 cm, but it can grow to 1 m. Its native range (Figure 1.5) is between the Hindu Kush and Novo Sibirsk in central Asia, in the foothills of the mountains, covering roughly 2500 km by 800 km (Hultén and Fries, 1986). It was brought to the Botanic Gardens of Geneva, Switzerland and Dresden, Germany in 1837. 'For many years it was a typical ruderal plant, occurring only in towns, gardens, parks and cemeteries. A few decades ago, however, it started to penetrate into woods, at the beginning only to badly degraded places, but later on it became established also in quite natural stands of deciduous forests' (Kornaś, 1990). This change is shown in Figure 1.6. Mostly it is found in damp shady places in its introduced range. My guess is that the apparent habitat shift in central Europe is more a reflection of dispersal than adaptation, a proposition that could now be tested by molecular techniques.

The third invader got its name *capensis* from a misapprehension that it came from the Cape of Good Hope, South Africa. In fact it is North American with a range about 3000 km from west to east and 2000 km from north to south, from Newfoundland to Saskatchewan to Oklahoma to Florida (Figure 1.5). It is known as jewelweed from its attractive orange flower, and grows to about 1.5 m. It is established in the south and east of England, and in small parts of Wales and northern France (Perrins, Fitter and Williamson, 1993; Fitter, 1978). Although it is the least widespread invader of the three (and with the largest natural range), it is strikingly more successful in England than the native *I. noli-tangere*.

I. noli-tangere is a declining species in England, most common in the north-west. But globally, *I. noli-tangere* is not only much the most widespread (Figure 1.5), but also forms a ring species, an allospecies, with *capensis*. *I. noli-tangere* is found from France across Europe, Russia and Siberia to Japan and Kamchatka. In America its range is from southern Alaska, to the state of Washington and the province of Saskatchewan (Hultén and Fries, 1986). The relationship of the two species in Canada is not clear, perhaps both because many botanists are unaware that some jewelweeds might be *noli-tangere*, and because both have cleistogamous flowers as well as open pollinated ones.

The distributions and successes of *noli-tangere* and *capensis* in England are notably different. Morphologically they are very similar (Figure 1.4). *I. capensis* has an orange flower with blotches of brown rather than a yellow one, the spur at the bottom of the flower is held at a different angle, and the leaves are smaller on average, with slightly fewer teeth. The genetics of these differences are apparently not known, but similar differences between other closely related species of plants are determined by rather few genes, a very small fraction of the genome (Gottlieb, 1984; Orr and Coyne, 1992). All this is consistent with the view that small genetic changes can, on occasion, have important ecological effects, and that this may be one reason why invasions are so hard to predict. This view will be examined in Chapter 6 (section 6.2).

The spread of the three invasive species in Britain can be mapped using first records from vice-counties (Perrins, Fitter and Williamson, 1993). In the mid-19th century, H. C. Watson found that the sizes of the British counties were unsatisfactorily variable for recording the distribution of British plants. He devised the vice-county system of dividing the larger counties and merging small ones with their neighbours, leaving the middle sized ones untouched. This allowed him to record distributions using the maps of the day, a tradition that continues to this day in the Botanical Society of the British Isles, who call their journal *Watsonia*.

The average size of a vice-county is about 2200 km^2, making them only a touch smaller than 50-km grid squares which are used nowadays to map European distributions. So even now vice-counties are a useful way to map distributions, and they have the additional advantage that records back to the middle of the 19th century and earlier can be used (Dandy, 1969).

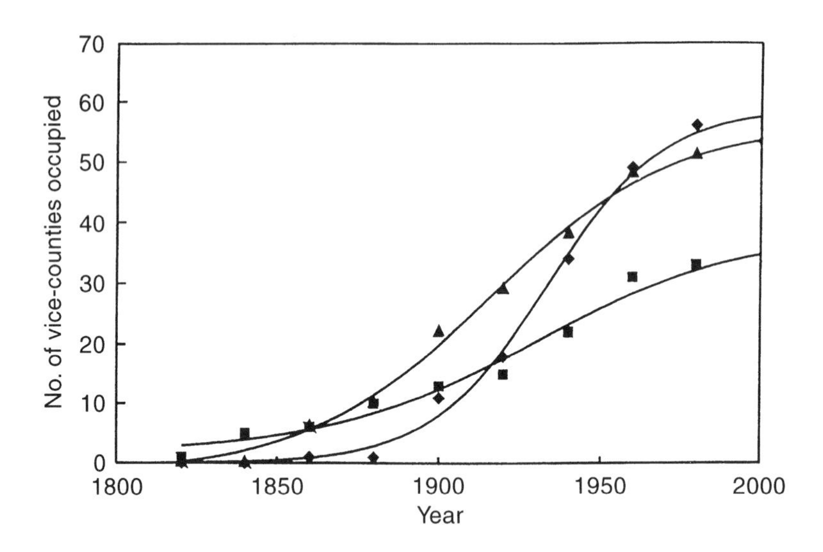

Figure 1.7 Cumulative records for the first occurrence of *Impatiens* spp. in English vice-counties, with logistic curves fitted to the data. ■, *I. capensis*; ▲, *I. parviflora*; ◆, *I. glandulifera*. (From Perrins, Fitter and Williamson, 1993.)

Unlike the spreading fronts seen in fulmars and rabbits (Figures 1.1 and 1.2), the spread of *Impatiens* is blotchy (Perrins, Fitter and Williamson, 1993), reflecting long-distance spread by human transport. Nevertheless, the cumulative records give smooth curves (Figure 1.7), which allow comparisons of the rate and extent of the spread of the three species. The data are fitted well by the well-known logistic curve, and with some simplifying assumptions the parameters of the fitted logistics can be manipulated to give some interesting comparative statistics (Table 1.3). (The logistic curve is discussed more fully in section 3.4.1 and in the box in Chapter 7.)

It can be seen that both *I. glandulifera* and *I. parviflora* can be expected to invade all or almost all the 59 English vice-counties, while *I. capensis* will probably invade less than two-thirds, which, from the maps, will be southern and eastern. Those figures come from the calculated asymptote, the K of the logistic. It is also possible to calculate the rate of increase at any point of the curve, and the position of the curve on the time axis. Assuming that all vice-counties are the same size and contiguous gives figures in kilometres per year, which will be underestimates because of the blotchy spread. In Table 1.3 these have been calculated for the initial spread (from one vice-county), and for the fastest spread, at the point of inflection of the logistic curve whose date is also given. *I. glandulifera* is the fastest and latest in its spread, *I. capensis* the first and the slowest. *I. glandulifera* is also the tallest, the most likely to form large mono-specific clumps, i.e. the most gregarious, with the highest seed

Table 1.3 Invasive *Impatiens* in Britain

	Species		
	glandulifera	*parviflora*	*capensis*
Imported	1839	?	?
First record in wild	1855	1848	1822
Initial rate of spread (km year^{-1})	2.6	1.6	1.4
Maximum rate of spread (km year^{-1})	38	24	13
Date of maximum rate	1933	1915	1928
Predicted final number of English vice-counties invaded	59	58	37

Data from Perrins, Fitter and Williamson (1993).

production and probably the most frost-resistant as a seedling (Perrins, Fitter and Williamson, 1993) which allows it to germinate early in the spring. All these characters probably help it spread and establish more readily then the other two species.

These *Impatiens* species again illustrate many of the conceptual framework points. All were brought in as garden plants (CFP 0), and all show how unpredictable invasions are from the ecology in the native habitat (CFP 4), the importance of propagule pressure (CFP 2) and the variation in habitats invaded (CFP 3). In these cases, spread by humans is three of four orders of magnitude faster than natural spread (2 to 20 km yr^{-1}, Table 1.3, rather than 2 m yr^{-1}) and is more irregular than the fulmar and rabbit (CFP 4). Their consequences are on the whole minor, but *I. glandulifera* can be a pest because of its swamping growth and early germination (CFPs 6 and 7). The small genetic but notable ecological difference between *I. capensis* and *I. noli-tangere* is instructive (CFP 8).

Now that these four examples have introduced the conceptual framework of Table 1.1, it is time to look at the points of that framework systematically and in more detail.

Appendix

THE SCOPE PROGRAMME ON THE ECOLOGY OF BIOLOGICAL INVASIONS

As this programme produced so much background information on invasion it may be helpful to list here first, the questions the programme tried to addres and second the publications that came from it. The list in Drake *et al.* (198 is, for timing and other reasons, incomplete. The programme ran for about years from 1982.

SCOPE is the Scientific Committee on Problems of the Environment, a international non-governmental organization, and part of ICSU, the Interna tional Council of Scientific Unions.

The SCOPE programme questions were:

1. What factors determine whether a species will become an invader or not
2. What site properties determine whether an ecological system will be pro to, or resistant to, invasions?
3. How should management systems be developed to best advantage, give the knowledge gained from attempting to answer questions 1 and 2?

PUBLICATIONS FROM THE SCOPE PROGRAMME

Brown, C.J., Macdonald, I.A.W. and Brown, S.E. (eds) (1985) Invasive alien organism in South West Africa/ Namibia. *South African National Scientific Programm Report. No. 119*, CSIRO, Pretoria.

di Castri, F., Hansen, A.J. and Debussche, M. (eds) (1990) *Biological invasions in Europ and the Mediterranean basin.* Kluwer Academic Publishers, Dordrecht.

Drake, J.A., Mooney, H.A., di Castri, F., Groves, R.H., Kruger, F.J., Rejmánek, M and Williamson, M. (eds) (1989) *Biological Invasions, A Global Perspective.* Joh Wiley and Sons, Chichester.

Gray, A.J., Crawley, M.J. and Edwards, P.J. (eds) (1987) *Colonization, Succession an Stability.* 26th Symposium of the British Ecological Society. Blackwell Scientifi Publications, Oxford.

Groves, R.H. and Burdon, J.J. (eds) (1986) *Ecology of Biological Invasions: A Australian Perspective.* Australian Academy of Science, Canberra.

Groves, R.H. and di Castri, F. (eds) (1991) *Biogeography of Mediterranean Invasion* Cambridge University Press, Cambridge.

Kornberg, H. and Williamson, M.H. (eds) (1987) *Quantitative Aspects of the Ecolog of Biological Invasions.* Royal Society, London.

'oenje, W., Bakker, K. and Vlijm, L. (eds) (1987) The ecology of biological invasions. *Proceedings of the Koninklijke Nederlandse Akademie van Wetenschappen*, **90**(1), 1–80.

Macdonald, I.A.W. and Jarman, M.L. (eds) (1984) Invasive alien organisms in the terrestrial ecosystems of the Fynbos biome, South Africa. *South African National Scientific Programmes Report No. 85*, CSIRO, Pretoria.

Macdonald, I.A.W. and Jarman, M.L. (eds) (1985) Invasive alien plants in the terrestrial ecosystems of Natal, South Africa. *South African National Scientific Programmes Report No. 118*, CSIRO, Pretoria.

Macdonald, I.A.W., Jarman, M.L. and Beeston, P. (eds) (1985) Management of invasive alien plants in the Fynbos biome. *South African National Scientific Programmes Report No. 111*, CSIRO, Pretoria.

Macdonald, I.A.W., Kruger, F.J. and Ferrar, A.A. (eds) (1986) *The Ecology and Management of Biological Invasions in Southern Africa.* Oxford University Press, Cape Town.

Mooney, H.A. and Drake, J.A. (eds) (1986) *Ecology of Biological Invasions of North America and Hawaii.* Ecological Studies 58. Springer-Verlag, New York.

Ramakrishnan, P.S. (ed) (1991) *Ecology of Biological Invasions in the Tropics.* International Scientific Publications, New Delhi.

Usher, M.B. (ed) (1988) Biological invasions of nature reserves. *Biological Conservation*, **44**(1,2), 1–135.

2 The origins and the success and failure of invasions

Conceptual framework points (from Table 1.1):

CFP 0 Most arrivals at present are from human importations, but natural arrivals are also of interest
CFP 1 Most invasions fail, only a limited number of taxa succeed (tens rule)
CFP 2 Invasion (or propagule) pressure is an important variable. So invasions are often to accessible habitats by transportable species

2.1 INTRODUCTION

Where are invasions most frequent, and why? It is often said that invasion happen more readily in disturbed sites than elsewhere. Certainly they are mor common there, and it might be thought that that tells us something about th biological nature of invasions. But it only reflects the fact that species are mor likely both to be transported from disturbed areas and to arrive in them because of human activities. At least, that is the case with terrestrial organism With invasions in the sea, there is no concentration on disturbance in the wor of marine scientists, more on the origin of invaders, on the mode of transpo and on the characters of the invaded place.

This chapter deals with these basic points about invasions: where do the come from, how do they travel, what are their chances of success? An following on from those: how often do invaders become pests, and how doe the pattern of introduction, i.e. the propagule pressure, affect what happen in invasions and our perception of them? In the Conceptual Framework (Tab 1.1) these are the first points, CFP 0, CFP 1 and CFP 2, and they are repeate at the head of this chapter. Everything else that can be said about invasions a a set depends on understanding these three points; they are the basis for th statistical study of invasions.

Two other points relate only to some invasions, but fall conveniently her These are the phenomenon of boom-and-bust, and the comparative behavio of invasions of closely related species.

2.2 THE ORIGIN OF INVADERS (CFP 0)

Invasions have been an important component of the evolutionary process throughout geological history. But nowadays, most invasions happen because of human activities, from commerce, agriculture and travel. Even though (as will be discussed in section 2.3) most invaders fail and have small effects, the cumulative effect of those that succeed has been, and will continue to be, large. Although the effects are strongest in agriculture, they affect all systems (CFP 3, section 3.2). As Charles Elton said 'We must make no mistake: we are seeing one of the great historical convulsions in the world's fauna and flora' (Elton, 1958 p.31).

Elton was particularly interested in invasions across Wallace's biogeographical realms (Figure 2.1), such as the rabbit from the Palearctic to the Australasia. This type of invasion has certainly been important in the evolutionary process. A well-known example is the Great American Interchange (Stehli and Webb, 1985; Webb, 1991), the progressive exchange of mammalian faunas, and their subsequent evolution, between North and South America. This was the result of tectonic movements that brought these continents close together and formed the Central American bridge between them, a process starting ten 10 million years or so ago in the Miocene, and possibly still underway. Similarly, there has been an exchange of mollusc species between the Atlantic and Pacific, via the Arctic Ocean, since the Bering Strait first opened about 3.5 million years ago (Vermeij, 1991b). These are discussed in the context of community effects of invasion in section 7.4.2.

Unlike these paleontological invasions, modern invasions are too recent to result in major evolutionary effects like speciation; the time scale is out by three or more orders of magnitude, as explained in more detail in section 6.4.3.

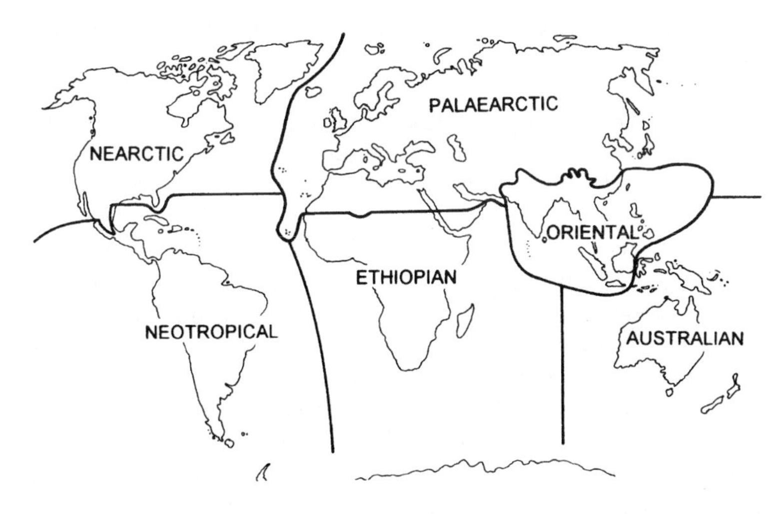

Figure 2.1 Wallace's six realms. Different authors use slightly different boundaries in some places. (From Wallace, 1876; illustration by Mike Hill.)

Minor evolution can be important, as shown by the rabbit and myxom
(section 1.3.2), and will be considered again in section 6.4.2. But in genera
modern invasions are an ecological rather than an evolutionary phenomeno

What a large ecological phenomenon they are! Heywood (1989) shows th
in modern floras, from a range of countries, at least 10% of the species ar
introduced, rising to almost 50% in New Zealand. For insects, Simberlo
(1986) shows a steady increase since the mid-19th century, reaching 1.7% c
the USA fauna by 1970, but 29% in Hawaii. The difference on islands
probably not so much a question of vulnerability (see section 2.3.2), but
combination of a statistical artefact (fewer native species), greater propo
tional change to the landscape, and a greater enthusiasm for deliberate intr
ductions in the past.

Because the number imported depends on what transport there is, th
number establishing reflects the ease of transit. Most lepidopteran larva
require living plant matter; many coleoptera do not. There have been mor
hymenoptera imported intentionally to the USA, usually for biological co
trol, than Diptera. World-wide these four orders constitute the bulk, thre
quarters or more, of the Insecta, roughly in the proportion two coleoptera t
one each of the other three (Groombridge, 1992). Yet up to 1980 the numb
of immigrants to the 48 contiguous states of the USA were Hymenoptera 39
Coleoptera 372, Lepidoptera 134, Diptera 95 and all others 692 (half of the
Homoptera) (Sailer, 1983). The probability of successful transit and delibera
introduction affects, in unpredictable ways, the numbers of species that arri
to invade.

In the 19th century, the pattern of colonization and trade meant th
introductions were predominantly from Europe (di Castri, 1989). Europea
settlers often set about recreating a European agricultural landscape, an
started acclimatization societies to introduce European species. Nowaday
the flow of commerce is much more widely spread, and faster, and speci
travel in all directions. One example of this is the flow of marine organisms i
ballast in ships. This happens on all the major shipping routes in the Pacif
(Carlton, 1987) and no doubt world-wide too.

Ships have used water as ballast regularly since the 1880s. Holds a
flooded for balance and stability, and then discharged at or near ports t
make room for new cargo. But since the advent of very large tanker
container ships and bulk cargo carriers in the 1960s the volume of wat
discharged has increased greatly, ships are faster and the ballast tanks ar
cleaner. More species can be and are transported. Any species of plankto
or of benthos that has a planktonic larva, may be caught up and carried lon
distances in ballast water. At least 365 species were found in 159 ships arrivin
in Coos Bay, Oregon from Japan (Carlton and Geller, 1993). While most c
those imported this way will not survive in the new environment, enough d
to produce a major invasion force, a pressure of propagules. A list of 4
successful invaders (see Table 2.1), all probably brought in in ballast betwee
1971 and 1990, covers all parts of the world and eight phyla (Carlton an

Table 2.1 Examples of invasions from organisms probably carried in ballast water. All have invaded since 1971

Taxon	*Example*	*Number of other examples in Carlton and Geller (1993)*
Dinoflagellata	*Gymnodinium catenatum*	2
Cnidaria Scyphozoa	*Phyllorhiza punctata*	0
Cnidaria Hydrozoa	*Cladonema uchidai*	0
Ctenophora	*Mnemiopsis leidyi*	0
Annelida Oligochaeta	*Teneridrilus mastix*	0
Annelida Polychaeta	*Marenzelleria viridis*	1
Crustacea Cladocera	*Bythotrephes cederstroemi*	0
Crustacea Mysidacea	*Neomysis japonica*	2
Crustacea Cumacea	*Nippoleucon hinumensis*	0
Crustacea Copepoda	*Centropages typicus*	8
Crustacea Brachyura	*Hemigraspus sanguineus*	1
Crustacea Caridea	*Hippolyte zostericola*	2
Mollusca Gastropoda	*Tritonia plebeia*	0
Mollusca Bivalvia	*Dreissena polymorpha*	6
Ectoprocta	*Membranipora membranacea*	0
Pisces	*Gymnocephalus cernuus*	7

Abstracted from Carlton and Geller (1993).

Geller, 1993). A particularly remarkable one is the Ctenophoran (comb jelly) *Mnemiopsis leidyi*, which travelled from the western Atlantic to the Black Sea, causing havoc there. Its biology and effects are detailed in section 5.3.1.

How often do imports lead to successful invasions, and how often do they become pests?

2.3 THE TENS RULE (CFP 1)

Once an invader has arrived, has been imported, it may get away, living outside cultivation or captivity, becoming feral. It may establish a new population. It may become a pest. Can any statistical generalizations be made about these stages? If so, what are the biological and other reasons for these generalizations? Answers to the first question are now reasonably well established. Answers to the second are, as yet, more vague and less satisfactory. Some preliminary answers to the second can be found by looking at variation in the answers to the first. So in this section, I will first consider (2.3.1) what the tens rule is, then (2.3.2) deviations from it. The reasons both for the tens rule and for the deviations from it are discussed in section 2.3.3.

One thing that became much clearer during the SCOPE programme on the ecology of biological invasions (see the Appendix to Chapter 1) is that it is unusual for the course of a particular invasion to be predictable. Gilpin (1990), reviewing the SCOPE synthesis (Drake *et al.*, 1989), said 'we are never going

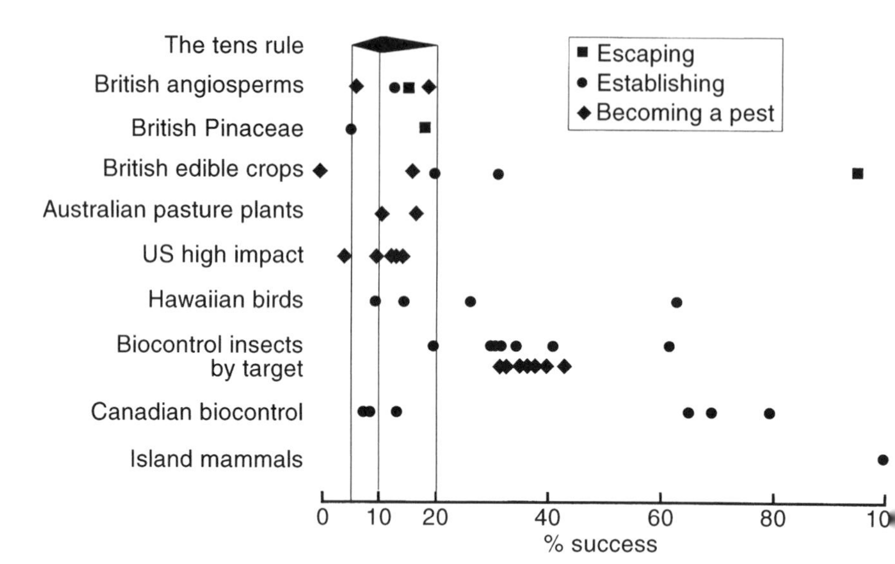

Figure 2.2 Ratios shown by invasions discussed in the text. The suggested range for th tens rule is shown at the top. Points fitting the tens rule fall between the vertical line Note the points outside the lines are mostly biological control and island example Different symbols, as shown in the box, are used for different sections of the rule. Th definitions of the stages are given in Table 2.2. British angiosperms from Table 2.3. Th two points for becoming a pest are for all pests and severe pests. British Pinaceae, se text. British edible crops from Table 2.4. Australian pasture plants are weedy species i Lonsdale (1994), legumes 10%, grasses 17%. US high-impact are, left to right, fishe plant pathogens, insects, molluscs, terrestrial vertebrates, from US Congress OT (1993; Figure 2-2). Hawaiian birds are, left to right, all species in native forest, all specie in native forest and open grassland, Columbiformes in all habitats, Passeriformes in a habitats (Table 2.5). Biocontrol insects by target from Table 2.6, with the symbol fo becoming a pest used for those effecting control (see text). Canadian biocontrol fror Beirne (1975) (see text). Island mammals from Ireland and Newfoundland combine (see text). (From Williamson and Fitter, in press, a; illustration by Jame Merryweather.)

to have a scheme to predict the success of invading species' and 'the study o invasion should be . . . statistical . . . characterizing the probability of outcome for classes of invasions'. This is an approach pursued where possible in thi book. It applies both to ordinary invasions and to releases of geneticall engineered organisms (section 6.5 and Williamson, 1993).

Figure 2.2. summarizes the data that will be discussed. The topic is th proportion of invasions going on from one stage to another; the stages ar defined below and in Table 2.2. There is a regularity in these proportions, a indicated in the top line of Figure 2.2, and a predictability about deviations That is the thrust of the argument in this section.

Table 2.2 Definitions

Imported	brought into the country, contained
Introduced	found in the wild, feral (common usage), casual, released but not (yet) breeding successfully
Established	with a self-sustaining population, naturalized, feral and breeding successfully, released and breeding successfully
Pest	with a negative economic effect
Escaping	transition from imported to introduced
Establishing	transition from introduced to established
Becoming a pest	transition from established to pest
The tens rule	each transition has a probability of around 10% (between 5% and 20%)

Enlarged from Williamson and Fitter in press (a, b).

2.3.1 The basic rule

How frequently do invasive species establish and how frequently do they become pests? A useful rule of thumb is the tens rule, that 10% of feral (or introduced) invaders, invasive species living outside captivity in any sense, become established, and 10% of those established become pests. For plants, 10% of those imported escape to become introduced (feral). Note the definitions in Table 2.2, and the discussion of them below.

This rule is useful in two quite distinct ways. On the one hand, as there are many exceptions, it is useful to have a yardstick, to be able to see if a particular class of invaders is unusually successful or particularly prone to failure. On the other hand, as the rule often holds, it is a major reference point in studying invasions. Success is fairly rare. The reasons for this are important in understanding the biology of invasions, and in taking appropriate action to minimize the nuisance from invaders.

The tens rule is a statistical rule, first put forward in 1986 as the ten-ten rule (Holdgate, 1986; Williamson and Brown, 1986) and extended to the third ten (from imported to introduced) in Williamson (1993). The 1986 version applied to a variety of British groups, vertebrates, insects and flowering plants. It seems to apply more widely than just to Britain. di Castri (1989) came up with effectively the same rule for plants in many temperate and Mediterranean areas. Other cases where the rule holds are shown in Figure 2.2 and discussed below.

Like all statistical rules, it needs to be interpreted with care. The tens in the rule are only roughly 10%. It is convenient to have a standard roughness, and this was put at 5 to 20% (Williamson, 1992). The roughness has to cover two things. One is the uncertainty, and reasonable disagreement, about the status of invaders, which I will come back to. The other reason is the inevitable sampling error.

If there are two classes, say established and non-established, with frequencies p and q $(p+q=1)$, then confidence limits can be found from the expansion of $(p+q)^n$. The confidence limits are not symmetric on the scale of p and q, so it is usual either to use tables or an angular transformation. The 95% confidence limits of 0.1 in a sample of 100 are 0.049 and 0.176, the 99% limits 0.038 and 0.20. The number 100 is arbitrary, though of the order seen in many data sets. These figures add plausibility to the suggested 5% to 20% range of the tens rule. It is easier to remember a doubling figure than exact confidence limits. It is best to think of the tens rule as meaning a small proportion, centring on around 10%, but with a definite but declining probability of being between 5 and 20%, as indicated in Figure 2.2.

While sampling error is inevitable, a much more major source of uncertainty and misunderstanding is the interpretation of the terms imported, introduced, established and pest. Concise definitions are given in Table 2.2. Imported covers species in collections or species accidentally brought into the country. Introduced means found outside control or captivity as a potentially self-sustaining population. Established means having a population of sufficient size and self-sufficiency that the species would persist indefinitely if the habitat is maintained. In dealing with plants, casual and naturalized perhaps best give the flavour of introduced and established. In the terms used in the regulation of genetically engineered organisms, introduced species are released, while imported but not introduced species are contained. The three transitions between the four stages are called escaping, establishing and becoming a pest (Table 2.2), and their probabilities constitute the tens rule.

There is much scope for disagreement about whether an imported species has become introduced, whether an introduced species has become established (see Table 2.4), and even more for whether a species is a pest, as is well known (Holdgate, 1986). There are no clear-cut lines between the categories. There is a range of views of which of a set of British annual plants are weeds (Perrins, Williamson and Fitter, 1992a,b); weeds are plant pests. Williamson (1993) gives a range of categories for British plants, and the number of species that two different experts think belong in various categories. Examples are 'severe pests', 'widely naturalized', 'fully naturalized', 'garden outcasts', and deciding which species fits which category is often somewhat subjective. All these uncertainties about the limits of classes inevitably blur the tens rule. Taking 10 to mean between 5 and 20 encompasses most of the possible, and perfectly reasonable, different views of which species to include in which category.

In Europe, Weeda (1987), Lohmeyer and Sukopp (1992), Sukopp and Sukopp (1993) and Kowarik (1995) are concerned with those established plant species that manage to establish populations in native vegetation. The transition from established anywhere to established in native vegetation is not covered by the tens rule. The figures are certainly different from any 'ten', being 52% in Germany and 37.5% in Netherlands, and are themselves significantly different ($P < 0.001$), but this probably reflects different views of what is native vegetation (Williamson, 1994). In Britain, almost all habitats are non-natural to some extent, even when most of the species are native; the same is true of the Netherlands and Germany. Hannah *et al.* (1994) class 98.7% of the British Isles as 'human dominated' and only 0.2% as 'undisturbed'. Kowarik (1995) distinguishes epekophytes which grow in communities that would disappear with the cessation of human activities, and agriophytes that grow in communities that would exist if human influence ceased. In places where there is much more undisturbed habitat, such as New Zealand with 27.1% (Hannah *et al.*, 1994), it is quite usual to call any invasive species of such a habitat a pest, irrespective of whether it has any measurable effect.

Kowarik (1995) also makes the excellent point that the proportions in the different classes will vary with time, because there can be a long period, often over 100 years, between a species being imported and becoming established. I return to one aspect of these time lags below when considering the 'boom-and-bust' phenomenon. Kowarik calls what I call imported introduced, which leads him to disagree with the ten : ten rule. But in fact he arrives at a 10 : 2 : 1 rule for the percent imported that become, in the terms here, introduced, established and agriophytes. I would write his rule as 10 : 20 : 50, while he might write mine as 10 : 1 : 0.1. On the first two transitions, escaping and establishing, we more or less agree. My last transition is becoming a pest, which he does not study, while his last one is becoming an agriophyte, which I do not study. His study shows, once again, the number of categories that could be distinguished in the continuum from imported to widely and fully established, and so the disagreements that can arise if the terms do not match Table 2.2.

With those cautionary remarks in mind, it is interesting to see both where else the tens rule holds and where it clearly does not (Figure 2.2). The latter is particularly useful in suggesting why the rule works at all, a point returned to in section 2.3.3.

Table 2.3 The tens rule* in British angiosperms

	Number	*Percentage*
Total imported	12507	
Total casual or more	1642	13.1
Total established	210	12.8
All pests†	39	18.6 (of 210)
Severe pests‡	14	6.7 (of 210)

From Williamson (1993).
*Ten means between five and twenty.
†Listed in Williamson and Brown (1986).
‡Listed in Williamson (1994).

The tens rule applies to British angiosperms (Table 2.3) and parts of it apply to various animal groups in Britain (Williamson and Brown, 1986). The rule also applies to British Pinaceae, where there are 116 imported, mostly growing in arboreta (Mitchell, 1982), 20 casual or introduced (Stace, 1991), mostly derived from those that are commonly used in plantations, and only one naturalized or established, which happens to be a distinct subspecies of the one native species (Scots pine, *Pinus sylvestris*). Each of those totals is cumulative, i.e. the imported total includes the introduced and established species. The tens rule predicts that one-tenth of a species will be a pest, so it is not surprising that there is not one. The plant figures are shown in Figure 2.2.

Lonsdale (1994) found that of 17% of 186 naturalized grass species and 10% of 277 species of legumes in the savannahs of Northern Australia were either weedy and useful or just weedy. That fits the pest part of the tens rule, and is shown in Figure 2.2. Williamson and Fitter (in press, a) took all Lonsdale's species as established, though he refers to them as introduced. All were imported to improve pastures, either extensively, sown into native pastures with minimal management, or intensively, sown in ley pasture cropping systems. As Lonsdale says 'previous studies have been hampered by uncertainties about the number of species that were introduced and failed to establish' but in his study 'this difficulty, at least, can be avoided'. Even so, it is possible that some would be classified as introduced rather than established using the categories in Table 2.2. Such a change would still keep the percentages within the tens rule. A remarkable feature of Lonsdale's study is that, although all these species were intended to be useful, only 6% of the grasses and 4% of the legumes are, and almost all these were weedy as well.

A survey of high-impact, non-indigenous species in the USA, covering terrestrial vertebrates, insects, fishes, molluscs and plant pathogens again show that around 10% of invaders are what might broadly be called pests. The percentages vary from 4% to 18% (US Congress OTA 1993, Figure 2-2). It is clear from the report that these are only the best estimates available at present. Even so, all fit the tens rule, or very nearly. Another case that fits the rule is the 53 actually or potentially disruptive species, out of 475 plant alien species, in Hawaii Volcanoes National Park (Taylor, 1992); that is 11.2% (not shown in Figure 2.2). There are a mere 250 known native plant species in the Park. Other cases that fit the tens rule come in the examples below where most of the ratios do not fit. Which are which is readily seen in Figure 2.2.

2.3.2 Deviations from the tens rule

There are a number of cases in which the tens rule, or parts of it, clearly do not hold (Figure 2.2), and these help to explain why the rule works when it does. Four will be discussed here: British edible crop plants, birds on the oceanic archipelago of Hawaii, insects used in biological control and mammals on two large continental islands, Newfoundland and Ireland.

Table 2.4 Fates of imported plants grown commercially for human consumption in Britain, with lists of some categories*

A	Total species imported	75	
B	Species found at least casual (introduced)	71	(95% of A)
C	Species possibly established (category D plus status of some doubt, e.g. very persistent)	22	(31% of B)
D	Species definitely established (naturalized)	14	(20% of B)
E	Pests in Canada (none is a pest in Britain)	3	(17% of C,D)

Names of plants in three of the categories, listed in the taxonomic order used by Stace (1991):

1 Not now found casual: four species

Phaseolus lunatus	Lima bean
Apium graveolens cultivars	Celery
Solanum melongena	Aubergine
Cichorium endivia	Endive

2 Intermediate between naturalized and casual: eight species

Brassica napus	Oil-seed rape
Raphanus sativus	Garden radish
Ribes rubrum	Red currant
Ribes nigrum	Black currant
Ribes uva-crispa	Gooseberry
Lycopersicon esculantum	Tomato
Solanum tuberosum	Potato
Helianthus tuberosus	Jerusalem artichoke

3 Naturalized: 14 species; those marked † are weeds in Crompton *et al.* (1988)

Humulus lupulus (cultivars)	Hop	Native non-cultivars
Morus nigra	Black mulberry	
Ficus carica	Fig	
Rumex acetosa ambiguus†	Common sorrel	Three native sub-species
Sinapis alba alba	White mustard	
Fragaria × *ananassa*	Garden strawberry	
Prunus cerasifera	Cherry plum	
Prunus domestica	Wild plum	
Prunus cerasus	Dwarf cherry	
Malus domestica	Apple	
Foeniculum vulgare	Fennel	
Pastinacea sativa hortensis†	Parsnip	*P. s. sylvestris* native
Cichorium intybus†	Chicory	'Possibly native'
Asparagus officinalis officinalis	Garden asparagus	*A. o. prostratus* native

*From Williamson and Fitter (in press, a).

(a) British edible crop plants

In discussions of risks from releases of genetically engineered organisms, it is quite common to find statements that crop plants will not become pests because they cannot grow outside cultivation (e.g. Brill, 1985; Caplan and Van Montagu, 1990). Such statements are largely not true (Table 2.4). Crop plants are almost all strongly selected to grow well where they are cultivated. Consequently, in the same regions, they normally grow well outside cultivation. In Britain, this means that almost all are either casual or naturalized, apart from seven species that are native anyway (and not included in Table 2.4). The escaping ten does not hold. There are 71 species casual or naturalized in Table 2.4. If the escaping ten held, there would be about 710 'not found casual', that is, confined to cultivation. In fact there are only four such, a most remarkable difference.

In Table 2.4, the proportion of established to casual is a little higher than would be expected from the tens rule, namely between 20 and 30%, depending on the view taken of the intermediates, such as the potato (Figure 2.2). From the rule, one or two species should be a pest. None is in Britain, but three are in Canada, a reasonable fit (Crompton *et al.*, 1988; Williamson, 1994).

(b) Hawaiian birds

The second case is Hawaiian birds (Table 2.5). Oceanic islands are well known to be vulnerable and invasible, much more so than continental islands (Elton, 1958; Williamson, 1981), so it is not surprising that over 50% of birds introduced have become established. But Table 2.5 contains two other important points.

Table 2.5 Introduced Hawaiian birds*

	Total introduced	*Established*		
		Anywhere	*In native forests*†	*In open upland*†
Passeriformes	51	33	8	2
Columbiformes	19	5	1	1
Total	70	38	9	3

*Data from Moulton and Pimm (1986); Pratt, Bruner and Berrett (1987); Simberloff and Boecklen (1991); Hawaii Audubon Society (1993); Moulton (1993).

†Species names (forest species first): *Cettia diphone*, Japanese bush-warbler; *Copsychus malabaricus*, White-rumped shama; *Garrulax pectoralis*, Greater necklaced laughing-thrush; *Garrulax caerulatus*, Grey-sided laughing-thrush; *Garrulax canorus*, Hwamei; *Leiothrix lutea*, Red-billed leiothrix; *Zosterops japonicus*, Japanese white-eye; *Cardinalis cardinalis*, Northern cardinal; *Streptopelia chinensis*, Spotted dove; *Alauda arvensis*, Eurasian skylark; *Tiaris olivacea*, Yellow-faced grassquit; *Pterocles exustus*, Chestnut-bellied sandgrouse.

The first is that the overall success rate is higher ($P<0.01$) for passeriformes (perching or song birds) than for columbiformes (pigeons and doves), as noted by Moulton (1993). It is not known why. The second point is perhaps more important. There is scarcely any native habitat left over most of the lowland in Hawaii, where most of the introduced species live. However, there are some invaders in the native forests, and a few in the high open habitats (Table 2.5). Both habitats have been much affected by other introductions (Stone and Stone, 1989), but taking the figures at face value, 11–17% (Figure 2.2) have established in native habitats (depending on whether or not the open upland ones are taken as native). Various figures from 5% to 20% can be found from subsets of the data in Table 2.5, but there are no significant differences between any of these subsets. So, surprisingly, invasion of native habitats by birds on Hawaii conforms to the tens rule. The high invasibility of the lowland areas would seem to come from the extreme change of habitat there.

(c) *Insects for biological control*

Part of the case relating to insects in biological control is given in Table 2.6, another part (Beirne's data) is discussed below. The two parts are shown separately in Figure 2.2. In the first part (Table 2.6) there are two sets: parasitoids used to control other insects (divided by the mode of life of the target species) and herbivorous insects to control weeds. Instead of a tens rule, there is now something closer to a threes rule. About one-third of the species introduced manage to establish, but there is significant heterogeneity between

Table 2.6 Biological control success by insects by target*

Target	*No. of introductions*	*Percentage established*	*Effecting control as % of established*
Parasitoids			
Root-feeders	148	19.6	34.5
Above-ground borers	610	28.5	30.5
Mixed exo/endo-phytic	196	30.6	40.0
Leaf-rollers/tiers	41	31.7	38.5
External folivores	293	35.5	36.5
Leaf-miners	89	41.6	43.2
Herbivores			
Weeds	1128	61.3	32.4
Mean		35.5	36.5
s.d.		13.2	4.4
Heterogeneity χ^2		$P<0.01$	NS

*Data from Lawton (1990) and Hawkins and Gross (1992). Table from Williamson and Fitter (in press, a).
NS, not significant.

the different groups (Table 2.6). Hawkins and Gross (1992) suggest that this difference is to do with refuges, that certain life-styles give better protection against enemies. Their hypothesis certainly also fits the weed data in Table 2.6 as weeds are generally more apparent than insects, and further suggests that the availability of accessible resources is one factor in the tens rule. The proportion giving some degree of control may perhaps be compared to the proportion of pests in the tens rule, as both involve an economic effect (Table 2.2). Note that there is no significant heterogeneity in the proportion controlled (Table 2.6). If it is apparency that is important, it acts by determining whether a population can persist, not what its ultimate size will be.

Considering that control agents are selected to be successful, it is surprising that only one-third succeed, and not surprising that this fraction is much larger than one-tenth. Williamson (1989a) and Lawton (1990) have drawn attention to a few papers that indicate that a major factor in establishment success of biological control insects is the size of the introduction. From Beirne's (1975) survey, increasing the number released from less than 5000 to more than 30 000 improves success from 9% (about that expected from the tens rule) to 79% (well above any average in Table 2.6). Similarly, increasing the number of individuals in a single release, from below to above 800, changes the percentage success from 15 to 65. Using more than ten releases rather than less gives 70% success compared with 10%. All these proportions are shown in Figure 2.2. Propagule pressure behind an invasion is an important determinant of its probability of success, and is discussed further below in section 2.4.2.

(d) Mammals on continental islands

The fourth case of deviations from the tens rule shows that, with a good ecological match, the probability of an invasive species establishing can be 100% rather than 10%. Both Ireland and Newfoundland were cut off by rising sea level soon after the last glaciation, when they were still covered with ice and tundra. The result is that both have strikingly impoverished boreal/temperate mammalian faunas. There are several species in both Britain and Canada that would be expected, from their distribution and general ecology, to thrive in Ireland and Newfoundland respectively, if they could only get there. For instance, there are no native species of voles in Ireland, compared with three in Britain. The absence of voles may explain the absence of tawny owls *Strix aluco* and short-eared owls *Asio flammeus* in Ireland. Bank voles *Clethrionomys glareolus* were introduced into south-west Ireland in the 1950s. They are well established and spreading steadily (Crichton, 1974). For Newfoundland, snowshoe hare *Lepus americanus* was introduced successfully in 1864, mink *Mustela vison* in 1938, the masked shrew *Sorex cinereus* in 1958 (to control larch sawfly *Pristiophora erichsonii*, an accidental introduction from Europe in around 1900) and red squirrel *Tamiasciurus hudsonicus* in 1963 (Dodds, 1983). With five examples from Ireland and Newfoundland we find

100% success. There is no record of failures on either island of species that would be expected on biogeographical grounds to succeed. The little owl, *Athene noctua*, in Britain is another example of this type of successful invader (see section 2.4.2).

2.3.3 Reasons for the tens rule

The second question put at the start of section 2.3 was what are the reasons for the tens rule and the deviations from it? In many studies of invasion three sets of factors seem to be important. The first is propagule pressure, the rate at which propagules, seeds, breeding individuals and so on are released. The second is the set of factors that allow species to survive, and increase, from low densities. The third is the set of factors that determine local abundance. Let us see how far these and other factors can explain both the tens rule and the deviations from it. The availability of sustainable resources, which is the factor that seems to emerge most strongly from the boom-and-bust phenomenon (section 2.4.1 below), comes in to both the second and third. I shall take each ten of the rule in turn. Propagule pressure is so important that I have made it CFP 2 and treat it in detail in section 2.4.2.

First then, becoming feral, or the escaping ten. The only evidence on the first ten of the rule is the contrast, in Britain, between angiosperms and Pinaceae, on the one hand, and a special subset of angiosperms, edible crop plants, on the other (Figure 2.2).

What explanations might there be? One is that many species are imported in small numbers to places from which they cannot invade. Plants with poor dispersal in botanic gardens are obvious examples. A second explanation is that although a species may thrive in captivity, and be widespread in, say, horticulture, it fails, for physiological reasons, to produce propagules either at all, or which get outside cultivated habitats. A third is that the propagules die or are eaten before the species is recorded. The evidence of the Pinaceae and of crop plants is that the more widely a species is propagated, the more likely it is to be recorded as introduced, as feral or casual. Given the extraordinary variety, and variety of origins, of plant species that are imported, it is not surprising that roughly only one in ten becomes introduced. For most groups of animals there will be no comparable figure.

The establishing stage is much the most interesting of the stages, of the tens. It is the transition from introduced (or casual) to established (or naturalized). Several phenomena are undoubtedly involved, others may be. The data sets are summarized in Figure 2.2. Those that fit this part of the rule are: Angiosperms and Pinaceae in Britain, birds in native habitats in Hawaii, and Beirne's data on small introductions of biological control insects (less than 5000 released in total, or less than 10 releases in all, or less than 800 in any one release). Rather more, 20–30%, of edible crop plants in Britain, and columbiform birds in Hawaii, establish. Most groups of biological control insects succeed in establishing populations in about 30–40% of introductions, but one group has less than 20% chance, another more than 60%, perhaps related to apparency. Generally successful, but not universally so, at 60–80% success, are passeriform birds in Hawaii, and Beirne's large introductions (more than 31 500 individuals in total, or more than ten attempts, or more than 800 in any one attempt). Finally, introductions that would be expected to succeed on biogeographical grounds, mammals in Newfoundland and Ireland have (on a small sample) a 100% probability of establishing.

What do these examples tell us of the factors involved? First, propagule pressure is clearly important. Second, the ability of a species to maintain itself until conditions are favourable, and to increase from rare, are probably important differentiators. Possibly the difference between columbiform and passeriform birds in Hawaii relates to these. However, the first, and probably the second, are not involved in the failure of many British edible crops to establish feral populations. More basic population dynamic effects must be involved, which may be subsumed under the generality of suitable habitat, a habitat favourable to all aspects of population dynamics.

The two basic processes of population dynamics are reproduction and death. To establish a feral population the first rate must exceed the second. Observation of British casual species suggests that a failure of reproduction in adults is a common phenomenon, but a high death rate in young stages of the next generation can be important. Climatic matching, in as far as it is important (see section 3.4.5) probably most commonly acts in this way, rather than by affecting the general health of adults.

However, a sustainable state requires sustainable resources and not too many enemies. The boom-and-bust phenomenon, discussed below in section 2.4.1, suggests that sustainable resources are frequently lacking. The study of British squirrels, the native red squirrel *Sciurus vulgaris* and the invasive grey squirrel *S. carolinensis*, discussed in section 4.5, shows that even when the population effect is obvious (Okubo *et al.*, 1989) the resource involved may be subtle (Kenward and Holm, 1993), a difference in the digestibility of acorns for the two species. Cornell and Hawkins (1993) show that invaders generally suffer a lower rate of attack from parasitoids than natives. A lack of enemies, whether predators, parasites or diseases, has often been suggested as a major reason for differential success. The data on biological control insects (Table 2.6) suggest that some broad classes of habitats, or resources, differ in a predictable way.

Many have hoped that establishment can be predicted from the existence of empty niches, or climatic matching. When these correspond, as for mammals in Newfoundland and Ireland, prediction is possible. Otherwise these factors are of rather little use in prediction, as will be seen in sections 3.4.5 and 3.4.6.

Becoming a pest, the final ten of the rule, the probability that an established or naturalized species becomes a pest is one in ten, is supported by three examples. These are angiosperms in Britain, pasture plants in northern Australia and imported pests in the United States. The only contrary case, and one of a rather different sort, is the transition from failure-to-control to effective-control in insects used for biological control. Here the probability is about one in three.

Pests are difficult to define (Holdgate, 1986) but nevertheless very important in the study of invasions. The cost they inflict is enormous (US Congress OTA, 1993), and the probability of a genetically engineered organism becoming a pest is widely discussed (Williamson, 1994). But all the indications are that they have no particular properties as species; each pest is a pest for its own reasons. This can be seen in the characters of British pest plants (Williamson, 1994) and in pest vertebrates (Ehrlich, 1989). Pests, compared with other invaders, are more suited to anthropogenic situations, and have high densities. So the 10% of established species that become pests, and the third of established biocontrol insects that produce detectable control, may just represent the upper tail of the log-normal distribution commonly found (at least approximately so) in many communities (Williamson, 1981). If so, studying why and how species establish may be a better path to understanding pest status than studying pests alone.

2.4 FEATURES OF SUCCESS AND FAILURE IN INVASION

There are three phenomena connected with success and failure that I have yet to discuss in detail. The first is boom-and-bust, which, although not common, may nevertheless help in understanding the course of invasions in general. A much more important phenomenon is that of propagule or invasion pressure, already referred to several times, which I will examine more systematically. The third subsection of this section is a discussion of how far related species behave similarly as invaders.

2.4.1 Boom-and-bust

Sometimes an invading species goes to a peak of density and then declines, a path often called boom-and-bust. This phenomenon may throw a little light on the tens rule. In Britain, cases of it are known but are unusual (Williamson and Brown, 1986). They include the little owl *Athene noctua* (see section 2.4.2), Canadian pondweed *Elodea canadensis* (Simpson, 1984), the rhododendron lace bug *Stephanites rhododendri* (Hemiptera, Tingidae), but none seems to have led to the extinction of the invader. Some aquatic cases from round the world are discussed later in the book: *Cichla ocellaris* and *Mnemiopsis leydii* in section 5.3.1, the goatfish *Upeneus moluccensis* and other Lessepsian (Suez canal) migrants in section 7.4.2.

Most rabbit introductions to islands produced permanent populations, but there have been boom-and-busts, for instance both on Laysan (section 1.3.2) and Lisianski, also in Hawaii (Flux, 1994). In Shetland (see Figure 1.1) rabbits died out on the island of Trondra (300 ha), and the much smaller Holm of Melby (having survived cats *Felis catus* and shooting), Mooa and Oxna, though there are many more such islands were they survived (Venables and

Venables, 1955). In four cases, on temperate islands between 50 and 120 ha in size, including South Havra in Shetland, cats brought in to control rabbits ate them all and then died out (Flux, 1994). It is perhaps surprising that they failed to survive on birds, but that may be a consequence of the over-population induced by rabbits and the small size of the islands. On larger islands such as the Monach Islands in the Outer Hebrides (Macdonald, 1991) and others listed by Flux (1994), cats introduced to control rabbits persisted. Cats, as an example of invaders that cause marked ecological damage, will be discussed in section 5.3.1.

The same phenomena are known in freshwater fish (Welcomme, 1992), if rarely. In 28 cases (one case is one species at one place) out of 972 in which an introduction became established, the species disappeared. Some of those that persisted 'notably *Lepomis cyanellus* [Centrachidae], *Oreochromis mossambicus* and *O. niloticus* [Cichlidae] showed the typical "boom-and-bust" cycle which might be expected of any species transported into waters where its preferred food is under-utilized by existing species and where the resulting population explosion is later brought into equilibrium with available resources.' In fish it would seem that a bust leading to extinction is more common than one that allows a continuing reduced population. Those that do persist are not that rare. Welcomme (1992) classes *L. cyanellus*, green sunfish, as a pest, and both *Oreochromis* species form dense, stunted populations. Fish managers view *O. mossambicus*, Mozambique tilapia, with mixed feelings, sometimes a pest, sometimes a benefit, *O. niloticus*, Nile tilapia, as generally beneficial. In section 5.2.2 there is a case showing that some fish managers have a very different standard of benefit than environmental scientists.

Welcomme (1992) studied 291 species, but I have not been able to use his data to examine the tens rule. The chief reasons are that unsuccessful introductions are badly under-reported, that the statistics are of cases (species × places) rather than of species, and that the species to be imported are generally selected, in the same way that biological control agents are.

Reindeer *Rangifer tarandus* on three islands in the Bering Sea show the same boom-and-bust pattern, though only on one, St Matthew, would the population have become extinct, being reduced to 41 females and one sterile male (Williamson, 1981). The cause is certain on St Matthew, and is probably the same on the other two. Reindeer feed preferentially on lichens. On St Matthew they changed a lichen sward to a grassy one, and died from starvation. In other parts of the world, both on islands and continents, reindeer have persisted after going through a single boom-and-bust cycle. Indeed, all introduced reindeer populations fall into three classes, failure without a boom, failure after a boom and success after a boom (Leader-Williams, 1988). There are no cases of a successful population of this species increasing monotonically from introduction to asymptote.

The extinctions may all be ascribed to unstable predator–prey cycles a mismatch between the reproduction and longevity of the consumer and the consumed. More broadly, these failures to establish sustainable populations can be blamed on a lack of availability of accessible resources, the same explanation as that suggested for the variability of success in establishment of biological control insects (Table 2.6). Conversely, a permanent population implies sufficient accessible and sustained resources, but the population path may involve oscillations.

Hawaiian passerines seem to have followed the boom-and-bust pattern frequently as judged by the time to extinction. Moulton (1993) gives introduction and extinction dates for those on the island of Oahu. Time to extinction varies from 1 year to 40, with a median of 12.4. Some successful introductions such as the mynah *Acridotheres tristis* (Stone, 1989) and the red-billed leiothrix *Leiothrix lutea* (Pratt, Bruner and Berrett, 1987) have also shown the pattern.

It would seem that simple and anthropogenic habitats are particularly prone to show this pattern, which is most plausibly generally ascribed to interactions between the invader and its prey species. Decline, and extinction, from a build-up of enemies (predators or pathogens) seems less likely. Competition, despite strong advocacy (Moulton 1993), seems the least likely explanation in Hawaii, or for most of the examples. A lack of sufficient resources to sustain a population looks likely to be an important explanation of failures of invading animals to establish permanent populations. It is at first sight a much less plausible explanation for a plant such as *Elodea*, but it could fit with Tilman's (1982) strongly argued views on the role of resource ratios in structuring plant communities. That is, the decline might have come not from the lack of any one resource but from the wrong ratios. For *Elodea* there is no evidence of what caused the boom-and-bust; we can only say that genetics was not involved, as it is another clonal species (Williamson, 1994), like *Veronica filiformis* (section 1.1).

2.4.2 Propagule pressure (CFP 2)

Increasing the number of propagules increases the chances of a species establishing for several reasons. A few individuals introduced may disperse and not meet up again and so fail to reproduce. With more individuals there is greater chance of some finding a suitable habitat, or avoiding difficulties from the weather, or of not bringing in parasites or pathogens. General predators or herbivores can easily destroy a small initial inoculum. I shall first discuss three genera of birds, where the record is good, and then look at some statistical studies that show the importance of propagule pressure.

(a) Little owl

The little owl, *Athene noctua*, is a widespread palaearctic bird, found from Portugal to northern China, from Arabia to Latvia (Harrison, 1982). It is closely related to the burrowing owl *A. cunicularia* of North America. In Scandinavia it only occurs in Denmark west of the Great Belt (the widest passage from the North Sea to the Baltic), i.e. in Jutland and the islands of

Fyn, Aerø and Langeland (Dybbro, 1976). It was not native in Britain, though it was known as a rare vagrant, and was introduced several times in the 19th century. Like the mammals of Newfoundland and Ireland discussed in section 2.3.2, its absence from Britain appeared to be from a failure to cross a water barrier, the English Channel.

Some of the initial introductions failed, for instance a release of five in Yorkshire in 1842, and others in Norfolk and Sussex in the 1870s (Sharrock, 1976). Forty birds released in Kent in the late 1870s and more in Northamptonshire in the late 1880s led to established populations. Further largish introductions and natural spread led to its being widespread in central and southern England by 1920 and in all of England and Wales, with a handful of breeders in Scotland by 1970 (Sharrock, 1976). It is still extremely rare and southern in Scotland (Gibbons, Reid and Chapman, 1993). A large enough inoculum seems to have been necessary for the introduction to succeed.

Little owls are also a boom-and-bust species, as was noted at the start of section 2.4.1. There was a widespread decline in the central and southern English populations in the 1930s. Since the 1960s, the population has been fluctuating, with the largest yearly populations about twice the smallest, but with no clear trend (Marchant *et al.*, 1990). Hard winters in the 1940s and 1960s, and pesticides in the 1950s, have been blamed for part of the decline (Gibbons, Reid and Chapman, 1993) but these factors do not explain either the lack of an increasing trend in the 1970s and 1980s, or, more particularly, the decline in the 1930s.

(b) Common starling

The second bird is the common starling, *Sturnus vulgaris*, a western palaearctic species. The other 15 species in the genus are palaearctic and oriental only, and none has been an invader. The common starling appears to have increased in western Europe in the 19th and 20th centuries (Sharrock, 1976). For instance, it colonized the island of Skokholm off south-west Wales in 1940, and Bookham Common Wood, near London, in 1959, and has persisted in both ever since (Williamson 1981, 1987), though there has been a general decline in Europe since about 1960 (Marchant *et al.*, 1990; Gibbons, Reid and Chapman, 1993). It has been introduced into North America, South Africa, Australia, New Zealand and various other islands (Long, 1981), but not always successfully. Releases in Quebec, Canada failed in 1875, 1889 and 1892, and in the USA in New York in 1872 and Portland, Oregon 1890. The present successful, maybe over-successful, North American population derives from 60 birds released in Central Park, New York in 1890 and a further 40 in 1891, by a club wishing to establish all the birds named in the works of Shakespeare (Long, 1981).

Again, it may be just the size of that inoculum that was critical. Dobson and May (1986) show that starlings in America have only one-third of the species of parasite found in European birds, though 'no advantage has been demonstrated to accrue from this diminished list of parasites'. However, the Central Park birds may possibly have been more free of parasites than the failed releases.

(c) *House sparrow*

The third bird is the house sparrow, *Passer domesticus*. Like the starling, it has been very successful in anthropogenic habitats, particularly near houses and farms, in America and in many other parts of the world. The American parasite burden is less, with half the number of species found in Europe, two-thirds of that half acquired in North America (Dobson and May, 1986). As for propagule pressure, the first release of eight pairs in 1851 in the New York area failed, that of about 50 pairs in the same area in 1852 succeeded. The American spread was hastened by further releases in several places (Summers-Smith, 1988). I will return to the sparrow and its relatives in section 2.4.3 below.

There are several data sets that show that increasing the number released has a steadily increasing effect on the probability of establishment. Beirne's (1975) data on the introduction of biological control insects to Canada was discussed in section 2.3.2 (Figure 2.2). On a much smaller scale, Panetta and Randall (1994) experimented with seed sowings of a South African plant invader of Australia, an annual, *Emex australis* (Polygonaceae). The sowings were in 15-cm rings, and between 1 and 16 seeds were sown in each ring. They found significant effects not only with the number of seeds sown, but also from year-to-year and with seed burial, showing the importance of timing and chance in invasion success (see section 2.5 below). That is a retrospective study; *E. australis* is an important weed in Australian wheat.

With birds, Dawson (1984) found a marked effect in Long's (1981) data for introductions to New Zealand. Figure 2.3 shows that the probability rises from about the tens rule with less than 10 introduced, to near certainty with over 1000 introduced. That does not, of course, allow for differences in the ease of getting large numbers to New Zealand, or the prior probability of success of different species. In Australia, Newsome and Noble (1986), again using Long's data but supplemented by other information, found a strong positive effect of propagule pressure on the success of introductions of foreign birds, though with translocations of native birds the effect was insignificant.

A more sophisticated analysis, of translocations not alien introductions, is by Griffith *et al.* (1989). For birds and mammals in Australia, Canada, Hawaii, New Zealand and the United States they found a reasonable fit to their data using a stepwise logistic regression. Concentrating on the number introduced, their model can be written as

$$p = 1/(1 + kn^{-x})$$

where p is the probability of success, n is the number released, k a function of several factors and x another constant. This equation is asymptotic to $p = 1$ at

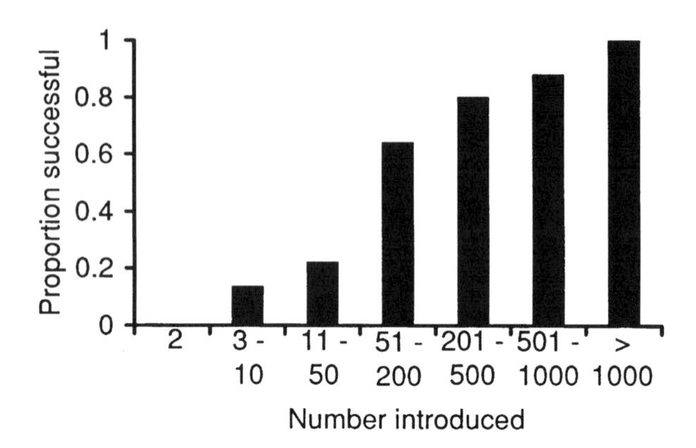

Figure 2.3 Proportional success of introducing different numbers of birds of a species in New Zealand. From Dawson (1984), based on 95 species, of which 41 succeeded, including three afterwards exterminated. Reintroductions, transplants and introductions of single birds excluded.

an infinite population, but the rate of approach is much affected by k. For instance, for a 3-year programme of reintroducing a threatened, endangered or sensitive bird species to the core of its historic range, the probability of success is quite low in good quality habitats, and far from certain even in excellent quality habitats. For the good habitats, the probability is only about 0.2 with 50 birds, rising to 0.3 with 400, while in excellent habitats the corresponding figures are 0.6 and 0.7. Numbers are important but so are other factors, factors that will be considered in section 3.3.3.

Two other studies show both the importance of propagule pressure and that the pressure comes from human introductions in most cases. In nature reserves both in southern Africa and in North America there is a striking positive linear relationship between the number of introduced species found and the logarithm of the annual number of visitors (Macdonald, Powrie and Siegfried 1986; Usher, 1988; Macdonald *et al.*, 1989). The effect is much stronger in North America than Africa and, in North America, stronger in woodlands than deserts.

A much older effect of human invasion is shown in Table 2.7. Carabid beetles, predatory ground beetles, can only cross large water gaps in ships. Scandinavian settlers unwittingly took these beetles, and many other organisms, with them as they voyaged to Faroe, Iceland and Greenland. It might be thought that Greenland's biota would be predominantly nearctic, from North America. In fact much of it, in southern Greenland, in many groups, is palearctic, from Europe. Only in northern Greenland, is the (very small) high arctic fauna Canadian and about as rich as would be expected (Downes, 1988). Southern Greenland behaves more or less as a boreal oceanic island, with an impoverished, disharmonic fauna; nearctic beetles have failed to reach it.

Table 2.7 Carabid beetles on North Atlantic islands*

	Shetland	*Faroe*	*Iceland*	*Greenland*
No. of species established	54	26	19	4
No. of species predicted from climate	56	53	54	23

*From Coope (1986).

2.4.3 Related species

While I agree with Daehler and Strong (1993) and also Lawton (1990) that previous success at invasion is a good indicator of whether an invader will succeed in a new place, I disagree that that is true for invasions by related species. *Impatiens capensis* is a successful invader in south-east England, but its allospecies *I. noli-tangere* (Hultén and Fries, 1986) is a rare and declining native species centred on north-west England (section 1.3.3) and the characters distinguishing the two would normally be called trivial. *Casurina equisetifolia* (called Australian pine but a dicot) is a pest in Florida, *C. glauca* persists only locally there (National Research Council, 1989). Brome grasses from Europe have very different patterns of success in America. *Bromus tectorum*, cheat grass, is a major pest in the inter-mountain west (Mack, 1989), *B. mollis* and *B. brizaeformis* are much less common there; *B. mollis* is more successful than *B. tectorum* in the central valley of California (National Research Council, 1989).

For vertebrates, Ehrlich (1986, 1989) notes several cases of differential success in closely related species, including the sparrows discussed below. His other cases are *Homo* versus other great apes, coyote *Canis latrans* versus wolf *C. lupus*, *Rattus* spp. (which will be discussed in section 5.3.1) and some reef fishes. *Cephalophilis guttatus* established after 2285 individuals were released on a Hawaiian reef, *C. urodelus* failed after more than 1800 were released. Both species are serranids from the Marquesas. *Lutjanus kasmira*, also from the Marquesas, established after 3100 individuals were introduced; *L. guttatus* from the Pacific coast of Central America, failed when 3400 were released. The differences in these reef fishes can obviously not be explained by propagule pressure. Taking the cases as a whole, small differences in biology and habitat preference seem to be the critical features.

Indeed, different genetic demes of a species may differ in their success at invasion. Some are discussed in section 6.2. There are many speculative such cases in the literature, such as rosebay willow-herb, or fireweed in North America, *Epilobium* (or *Chamerion* or *Chamaenerion*) *angustifolium*. This plant changed from a rare British native to an aggressive and widespread species early this century. Maybe new genes were introduced from garden or North American stocks (Salisbury, 1961), but the evidence, as in most such examples, is lacking.

There are several examples in this book where one member of a genus has spread and the others have not e.g. the common starling (section 2.4.2), collared dove (section 6.2), insects used to control *Salvinia* (section 5.4) and the *Impatiens* and *Rattus* species mentioned above. Sometimes it may be that only the one species is closely adapted to European style agriculture, but the case is usually not proved. Sometimes the other species in the genus have not been introduced, so (CFP 0) it is not known whether they would spread or not. Sparrows offer a rare case showing the importance both of genotype and of habitat.

Passer is a palearctic, oriental and Ethiopian genus of 20 species (Howard and Moore, 1991). The family is sometimes called Passeridae, sometimes Ploceidae. The world-wide success of the house sparrow, *P. domesticus* was mentioned in section 2.4.2. The Mediterranean equivalent and close relative of the house sparrow is the Spanish or willow sparrow, *P. hispaniolensis* (Figure 2.4). When they meet, they sometimes hybridize, as in Tunisia, sometimes behave as good species, as in Morocco (Mayr, 1963; Harrison, 1982; Summers-Smith, 1988). In north and central Italy, in Crete, Sicily and Corsica, there are stable, intermediate forms usually regarded as subspecies of *P. domesticus*, though Summers-Smith (1988) put them in *P. hispaniolensis*. These are probably of hybrid origin. In Spain and North Africa, when *P. domesticus* and *P. hispaniolensis* overlap, the house sparrow is the bird of houses, gardens and more intensive cultivation; the Spanish sparrow is in habitats bordering cultivation, in olive groves, scrub and woodland.

Out in the Atlantic, on Madeira, the only native passerid is the rock sparrow *Petronia petronia*. It was and is common throughout the rural parts of the island. In the 19th century it was also the town sparrow of the island. In 1935, the Spanish sparrow arrived, after 9 days of strong easterly winds, and rapidly became the only town sparrow (Bannerman and Bannerman, 1965). The Spanish sparrow is also the town sparrow in the Canary Islands and even further south in the Cape Verde Islands, where it has displaced the native Iago sparrow, *Passer motitensis iagoensis* (an African species) from the towns but not the country (Bannerman and Bannerman, 1968). A mixture of climate and habitat would seem to determine which sparrow succeeds where. Both the Canary and Cape Verde archipelagos were invaded in the 19th century. To complicate things further, there is a population *P. domesticus* derived from European not African stock, in Mindelo on the island of São Vincente in the Cape Verdes, which has been there without spreading since about 1922 (Summers-Smith, 1988).

The other *Passer* sparrow to be introduced widely is the (Eurasian or European) tree sparrow, *P. montanus* (Figure 2.4). (The American tree sparrow, *Spizella arborea*, is in a different family, Emberizidae.) As a native in northern Europe, *P. montanus* is not a town sparrow but more a bird of cultivated land with trees, park-land, orchards and so on. However, in China, Japan and south-east Asia, where *P. domesticus* is not found, *P. montanus* is a town sparrow, found in built-up centres, as well as in suburbs, villages and the surrounding cultivated land. It is also the town sparrow in Naples, Italy,

Figure 2.4 Three species of *Passer*: House sparrow, *P. domesticus*; Spanish sparrow, *P. hispaniolensis*; tree sparrow, *P. montanus*. Each is about 14 cm long. (From Harrison, 1982, with permission from HarperCollins Publishers.)

where the country sparrow is *P. hispaniolensis*. In Sardinia, *P. montanus* has invaded towns and villages along the east coast, starting around 1900, but probably spreading mostly in the 1970s (Summers-Smith, 1988).

When *P. montanus* meets *P. domesticus* the outcome varies. In northern Europe, as I said above, *P. domesticus* is the town sparrow. The same is true at the other end of *P. domesticus*' range, in Burma. But in central Asia, in Afghanistan and Turkmenistan, it is *P. montanus* that is the town sparrow (Summers-Smith, 1988).

In the United States, where *P. montanus* was introduced from Germany, it is spreading very slowly, in the order of 1 kilometre a year, from a park in St Louis, Missouri, in that state and across the Mississippi in Illinois. It has occupied about 20 000 km^2 after over a century (Ehrlich, 1986; St Louis and Barlow, 1988). This is a part of America where *P. domesticus* is abundant (North American Breeding Bird Survey, internet http://www.im.nbs.gov/bbs/bbs.html). The rate of spread of the house sparrow is considered in section 4.3.4.

In Australia, *P. domesticus* has been introduced from Europe and is abundant in much of the south and east and in Tasmania. It occurs almost throughout the eastern half of the island-continent. The rate of spread in different places varied from 6 to over 100 km yr^{-1} (Blakers *et al.*, 1984). Not all releases led to populations being established. That in Brisbane, Queensland died out, though house sparrows are now well established there having spread from further south (Blakers *et al.*, 1984). It is to some extent both a town and country bird in Australia. Hobbs (1961) writing about New South Wales near the Murray river said 'A common bird throughout the area, occurring regularly in the "bush" areas far from human habitations, as well as in the towns and around homesteads.'

P. montanus has also been introduced. The Australian tree sparrows are said to have come from China (Blakers *et al.*, 1984), probably southern China, though there is a possibility that the ancestry is partly European (Summers-Smith, 1988). The tree sparrow has been considerably more successful in Australia than in America, but still less successful than the house sparrow. It has spread from near Melbourne through parts of Victoria and into New South Wales. The average rate of spread appears to be about 5 km yr^{-1}, from the information in Blakers *et al.* (1984). It is generally only about a third as common as the house sparrow when both occur. However, at Waga Waga, New South Wales (about 150 km west of Canberra) there were 29.2 birds ha^{-1} compared with 35.1 birds ha^{-1} for the house sparrow, not a large difference (Blakers *et al.*, 1984), and it is slightly the more common in some areas.

The tree sparrow in Australia lives around towns in agricultural habitats, but so of course it does in China. Its spread has been slower than that of the house sparrow (which can use human transport), but faster than the tree sparrow in America. It has been recorded breeding only in 14 1° squares; the house sparrow breeds in 185.

The tree sparrow, presumably birds from Europe, failed in New Zealand (both North and South Island) and in Bermuda. Asian birds have successfully invaded several islands in south-east Asia (Long, 1981) and also towns and villages on tropical Pacific islands (Guam, Saipan, Tinian and Rota in the Marianas, Kwajalein and Yap in the Marshalls) (Pratt, Bruner and Berrett, 1987; not shown in Long, 1981).

All this shows that related species can behave similarly but also rather differently. In the final outcome, the differences may be more important than the similarities, yet the similarities allow some prediction of a range of possible outcomes. The more anthropophilic a bird is, the more likely it is to be introduced and succeed around human habitation, but that tells us nothing about invasion in general. House sparrows have been widely successful, though there have been a few failures probably ascribable to insufficient numbers being introduced, to a lack of propagule pressure. Where tree sparrows have behaved as town sparrows, in the Pacific and on Sardinia, they have been successful, though the spread has been slow compared with house sparrows. As country birds in agricultural areas, they have been much more successful in Australia than in America, possibly reflecting their geographical origin. Small genetic differences may be important in determining the course and pattern of invasion, a point I return to in Chapter 6, but the tree sparrow indicates some of the difficulties in testing such an idea. The variable success and rate of spread of this species is perhaps related to its surprising geographical distribution and variation.

2.5 CONCLUSION

Although we can hardly ever predict the success of individual invaders, there is no doubt that there are statistical regularities to invasions. The tens rule encapsulates much of what is known, and the deviations from it are illuminating. Propagule pressure and population dynamics are both involved, and both deserve better quantification. What else can be said?

Success and failure in invasion are two opposite facets. The reasons for success, CFP 4, are discussed in the next chapter. This chapter, and particularly the tens rule, emphasizes failure. Before turning to success, what is known of the causes of failure? As in any other enterprise, failure can come from bad luck or bad management. Among what can be regarded metaphorically as bad management are failures from a lack of resources or from an abundance of enemies. Failures from bad luck include the wrong genes or the wrong timing (Crawley, 1989).

Lack of sustainable resources comes in many forms. They are clearly implicated in some cases of boom-and-bust (section 2.4.1), and in the failure, from a lack of fleas, for myxoma to persist in the rabbits of Skokholm (section 1.3.2). A different sort of example is shown with measles on islands (Figure 2.5). Measles is a virus disease primarily of children. Once recovered, people are immune for the rest of their lives. So, the critical resource for measles is the number of susceptible school children. Before vaccination was introduced, the start of school terms often led to epidemics. The number of school children is closely related to the total population size. Mathematical models of the disease indicate that there is a minimum population for the disease to remain endemic (not to die out in a community), and for the disease to spread at all. Both these theoretical limits are shown in Figure 2.5, and they fit the data very well.

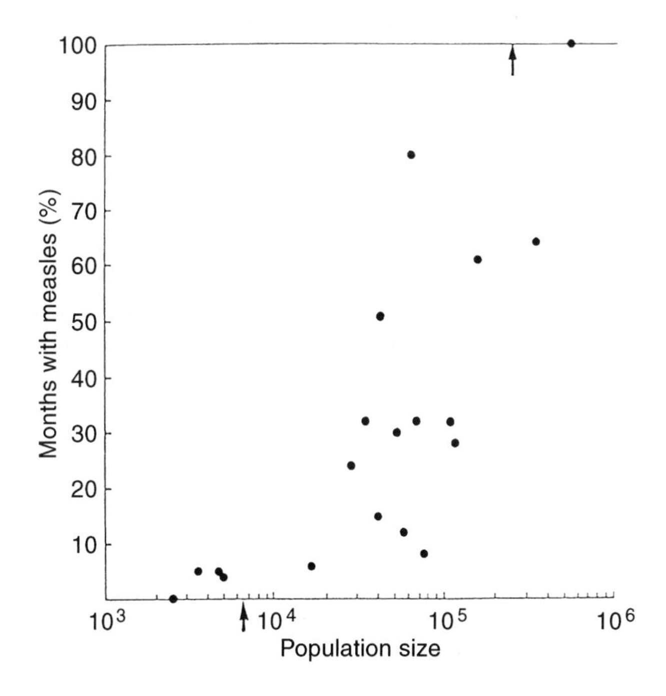

Figure 2.5 Extinction of measles in small island communities. Data from Black (1966). The upper arrow is Bartlett's (1957) estimate of the minimum population size for endemic measles. The lower arrow is Anderson and May's (1986) estimate of the minimum population size for measles to invade. (From Williamson, 1989b.)

Measles is one of the few cases where a mathematical population model can be used successfully to predict the degree of success. The general mathematical problem of modelling invasion patterns in communities will be considered in Chapter 7.

The importance of an abundance of enemies in preventing invasion, or conversely of an absence of enemies in allowing invasion, has often been discussed. Much of the evidence is anecdotal, but the record of biological control shows that sometimes an enemy species can severely depress the population of its victim. The relative lack of parasites, as we saw, may be important, but the point is difficult to prove. Many Australian plants have done well in South Africa and vice-versa, and quite often appear to have been imported with few of their native enemies. For instance, *Acacia longifolia* (Leguminosae or Fabaceae), a tree, produces 5.6 viable seeds per m^2 in its native Australia but 7370 per m^2 in Cape Province; *Hakea gibbosa* (Proteaceae), a shrub, has four times as many seeds per plant in South Africa as in Australia (Richardson *et al.*, 1992). In Australian sand dunes *Chrysanthemoides monilifera* (Compositae or Asteraceae) produces 2500 seeds per m^2, of which 2000 are viable. In its native South Africa the figures are 2300 (essentially the same) and only 50, a marked reduction (Weiss and Milton, 1984). But it is difficult to know how important enemies are in stopping an initial invasion. The evidence is more that a lack of enemies is sometimes important in an invader becoming a pest.

In the category of bad luck, the timing of the introduction of myxoma in Australia in relation to rainfall, and so sufficient mosquitoes to be a vector (section 1.3.2), is a simple example. Having the wrong, or at any rate different, genes has been suggested for rabbits (wild versus domesticated), fulmars (temperate versus arctic), *Impatiens* (*capensis* versus *noli-tangere*) and tree sparrows (German in America versus Chinese in Australia). In general the influence of propagule pressure, and particularly the statistical relationships, suggest that many invasions only succeed when they manage to find a window of opportunity (Richardson *et al.*, 1992).

Failure can come from a myriad of reasons. Is it easier to say what habitat and biotic factors go with success?

3 Which communities are invaded by which type of species?

Conceptual framework points (from Table 1.1):

CFP 3 All communities are invasible, perhaps some more than others

CFP 4 The *a priori* obvious is often irrelevant to invasion success. Among factors to consider: *r* (intrinsic rate of natural increase), abundance in native habitat, taxonomic isolation, climatic and habitat matching, vacant niche

3.1 INTRODUCTION

There are many misconceptions about invasions. Much is written that is at best half truth. But there are useful generalizations that can be made. This chapter deals with several factors that have been said to relate to the success of invasions; some do. Invasions can be studied in so many ways that I can only discuss topics that I think may be important. In section 3.3.3, in particular, it will be seen that the number of factors that may affect invasion success is very large. Some not treated here may turn out to have general importance. We still have much to learn.

3.2 WHICH COMMUNITIES ARE INVADED? (CFP 3)

Invaders are widespread but far from universal. They are common in disturbed areas; recent invaders are often rare in pristine communities. However, much of this variation comes from the variation in propagule pressure discussed in section 2.4.2. Looking for real differences in invasibility requires looking at the residuals from the relationship between invasion success and propagule pressure. As the form of that relationship is usually not known accurately, very little can be deduced about other effects. In general, there is too little reliable quantitative information to answer the second SCOPE question (see the Appendix to Chapter 1), 'What site properties determine whether an ecological system will be prone to, or resistant to, invasions?'

In as far as there is an answer, it is that all communities are more or less equally invasible. The apparent differences come from propagule pressure and the variation in the availability of suitable species. Impoverished communities have been thought to be more invasible, but they often occur in biologically difficult environments, such as salt-marshes or mountain tops. Any extra invasibility from a simplified community structure will be counteracted by the scarcity of species that can survive. Conversely, it has often been suggested that complex tropical communities are less invasible. This seems not to be so (Ramakrishnan, 1991), possibly because of the very large pool of potential invaders.

There is disagreement about how often invaders of anthropogenic habitats get into natural ones. An example of a species doing this is *Impatiens parviflora* in western Europe (see Figure 1.6). Rejmánek (1989) produced a list of only 60 species of plants world-wide invading natural vegetation, but Weeda (1987) found 75 species 'penetrating into natural vegetation' in the Netherlands alone. As noted in section 2.3.1, part of the difference no doubt comes from different views of what is natural vegetation. Some of these populations in natural communities may be sustained by immigration from less natural, disturbed, areas. Because of propagule pressure, most plant invaders on land are species of disturbed sites (section 2.1).

There are two studies which have looked quantitatively at the success of invaders in different ecological communities. Usher *et al.* (1988) looked at invasions into nature reserves world-wide and Crawley (1987) at invasions into different British plant communities. Both studies concluded that all ecological communities are invasible.

In Usher's Working Group on Nature Reserves, 23 reserves were studied, two on temperate islands, four on tropical ones, seven in mainland Mediterranean-type ecosystems, five in tropical savannahs and dry woodlands and five in arid lands. There were reserves in North and South America, Europe, Africa, Australia, Indonesia and on oceanic islands. The Working Group considered both plants and animals. Their two general conclusions were that there had been invasions into all the reserves, and that the strongest correlate was with the number of human visitors, probably a measure of propagule pressure (see section 2.4.2). In some subsets there were positive relations with area (bigger reserves got more invaders) and some islands had more invaders as a proportion of the native biota. The latter probably reflects the well known impoverishment of isolated islands (Williamson, 1981).

In Britain, Crawley (1987) counted the proportion of invasive vascular plant species noted in a standard flora in different habitats. Not surprisingly, he found high proportions in man-made habitats (waste, walls, fields and hedgerow). He found no invaders recorded in some habitats: upland summits, rock ledges and screes; the sea and brackish water; and oligotrophic lakes. All these are difficult habitats for vascular plants, and so species-poor. Any invader would have to be unusual. Alpine and rock gardeners might, for instance, grow plants that could invade upland rock ledges, but the plants are unlikely to reach them. Crawley's conclusion that all British plant communities are invasible is reasonable. The variation in invasion seen is largely explicable by propagule pressure. Lacking any direct measure of that pressure, it is impossible to say if there are other effects in the data.

While it is likely that some communities are more invasible than others, this has yet to be proved. There are wide differences in the rates of invasion; for instance, disturbed habitats have many invaders, but there is no good evidence that that relates to invasibility.

3.3 GENERALIZATIONS ABOUT INVADERS

If the answer to SCOPE's second question is that all communities are invasible (but some probably more than others), what can be said in answer to SCOPE's first question (see the Appendix, Chapter 1), 'What factors determine whether a species will become an invader or not?'. Here a wide variety of answers has been suggested. In section 3.4, I shall deal at some length with certain factors that have received a lot of attention. This section gives an overview and concentrates on two topics. The first (section 3.3.2) is the distinction between invader, colonizer and weed (or pest). Failing to distinguish clearly between these causes confusion and misunderstanding. At the same time attempts to determine the characters of weeds indicate how likely it is that it will be possible to find general characters for invaders. Secondly (section 3.3.3), there have been several interesting attempts to identify the characters of successful invaders by statistical methods. These show the wide variety of characters and considerations that may be important, indicate that different characters are important in different places and show quantitatively the relative importance of different characters. Generalizations can be made, but they need to be made about carefully defined and delimited situations.

3.3.1 Generalizations from biological control

A related area where there has been a long search for generalizations is in biological control. If effective rules could be found about which biological control agents will be effective, a great deal of time and money could be saved, and the whole technique would become more efficient. Biological control programmes have at least three goals, but only the first is identical with that of invasion biologists: what determines whether a species will establish? The second, whether an established agent will give effective control, is a separate question, as was seen in section 2.3.2. There is a rough correspondence between effective control and an invader becoming a pest. Whether that will happen is certainly of interest to invasion biologists, but it would be too optimistic to expect biological control work to be more than a weak guide here. The third question for biological control is whether the agent will be specific. An agent that attacks what are called non-target organisms, the wrong organisms, should be, and usually is, unacceptable (see section 5.2.1). A nasty example of what can go wrong with non-specific biocontrol is given in section 5.6. The frustrating point for those working commercially, and the interesting point in relation to invasions, is that less-specific organisms can often be introduced more easily. That is probably not so much that they can find alternative resources, though that is true, as that they are more ecologically flexible, for either genetic or physiological reasons or both, and so can adapt more easily to a new habitat.

Goeden (1983) developed a table of scores for different characters to try to predict success in one sector of biological control, the use of insects to control weeds, particularly agricultural weeds. The scores are an example of how applied entomologists assess the relative usefulness of characters or factors. Some of the factors that are relevant to establishment are topics discussed fully in section 3.4; many generations per year and many progeny in 3.4.1, large native range in 3.4.3, climatic matching in 3.4.5. Each of those gets a maximum of six points in Goeden's table. Others, such as a prolonged attack season (6 points), ease of culture (4 points) and gregarious feeding (3 points) are more probably related to control than to establishment. The role of enemies in the native range (maximum 6 points), mentioned in section 2.5, comes again in 3.4.3.

The highest scoring single factor in Goeden's table is previous success. 'Controls host in native home or one region of introduction' gets 6 points, but 'successful in two or more regions of the world' gets 12. Daehler and Strong (1993), among others, also give 'past invasion success by the potential invader' as a good indicator (see section 2.4.3). The importance of this factor suggests that it is the detailed biology of the invader that is more important than the details of the invaded community, though obviously the resources must be there.

3.3.2 Invaders, colonizers and weeds

If invasion is to be understood, it is important to distinguish invasion from colonization and weediness. These are three distinct but overlapping categories (Figure 3.1), and I define them in the next paragraph. This important distinction is sometimes difficult to disentangle in the literature, though the usage here is also followed by Rejmánek and Richardson (in press). Colonization is sometimes used to mean invasion, sometimes in the sense defined below, sometimes in a mixture of the two. Some distinguished scientists, such as Heywood (1989) and Newsome and Noble (1986) discuss plant invaders as if all successful ones were weeds, and so examine the characters of alien weeds to learn about invasion. But colonizers, weeds and invaders are different.

Invaders are species coming from elsewhere. The definition of invader is geographical. Lists of invaders will include the established species as defined in Table 2.2, and may also include the introduced category as defined there. Weeds, or pests, are simply organisms that somebody would like removed. In Table 2.2 they are defined as having a negative economic effect. That effect may be direct, causing damage, or indirect, by interfering with other human

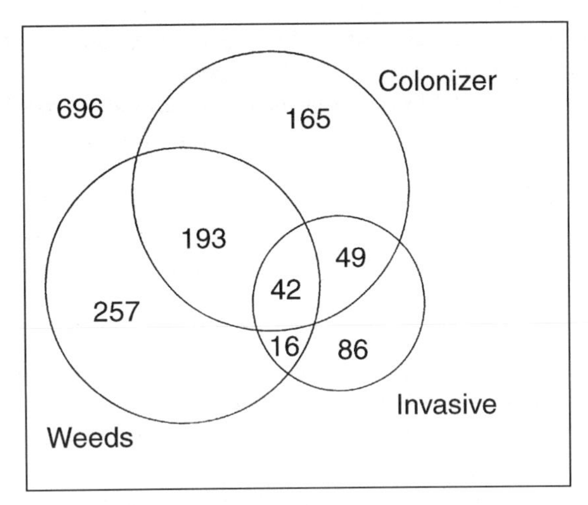

Figure 3.1 A Venn diagram showing the overlap of the sets of weed, invader and colonizer species in the British flora. The numbers in each sector are the species counts by Dr J. M. Perrins; the number 696 is the number of species that fall into none of the three categories. (From Williamson, 1993, and from Dr Perrins' D.Phil. Thesis, University of York, UK, 1991, with permission.)

activities. Weeds may be native or invasive. The definition of weed relates to human perception, and human perceptions vary (Perrins, Williamson and Fitter, 1992a). Colonists, in the sense used here (and also for instance by Gray (1986) and Grubb (1987)) are species whose ecological style is to keep moving to fresh territory, species whose population in one small place is not permanent. These species persist by perpetually founding new populations while losing old ones. Such species are particularly found in disturbed habitats. Like weeds they may be either invaders or natives. They are the only one of the three categories defined by biological characteristics. That in itself suggests that it may well be difficult to find any biological consistency in the characters of either pests or invaders.

As these three categories are distinct, they will have distinct characteristics. In Figure 3.1 there are eight sets, from the different combinations of the three categories. Although in the figure they are shown as clearly distinct, in practice this is not so, as we saw in the discussion of pests in the tens rule (section 2.3.3). Each of the boundaries in Figure 3.1 is, in reality, fuzzy.

The characters of invaders will emerge during this chapter. The characters of weeds are relevant to the general study of invasions, if only because it would be useful to be able to predict which invaders will become weeds. The characters of invaders that become weeds might reasonably be expected to combine the characters of both invaders and weeds, but that is only possible if both invaders and weeds have definable characters. Unfortunately, as will be seen, neither set as a whole seems to have a single set of characters. So, before considering the possible properties of successful invaders, consider the properties that have been associated with weeds, and, as a related topic, the successful biological control of weeds by insects.

Baker (1965, 1974) produced a famous list of the characters of the 'Ideal (?) weed', covering both annuals and perennials. His list includes characters such as seed production and growth rate, which are characters of the individual plant. Newsome and Noble (1986) defined the population styles of weeds. This approach perhaps reflects more realistically the variety of morphological and demographic features that allows a plant to become a weed, or an invader. For plants in Victoria, Australia, they identify Gap-grabbers, Competitors, Survivors and Swampers as four, largely distinct, sets of weedy invaders. The names define the ecological behaviours, though these derive from the characters of the individual plant. Each set has different characters. On this view, invaders are generally specialists in some sense; Baker on the contrary regards weeds as super generalists.

As Baker's characters are routinely used by agricultural companies to argue that their genetically engineered products will not become weeds (see section 6.5), it is worth examining these characters in more detail. There are two points to make. The first is that as Newsome and Noble are obviously right in distinguishing different classes of weed, no one set of characters can possibly predict weediness correctly. The other is that a statistical study of Baker characters in a restricted set of weeds produces a result that suggests that an intermediate number of Baker characters is associated with maximum weediness. The result in summarized in Figure 3.2.

To produce Figure 3.2 it was necessary to get a weediness score and a

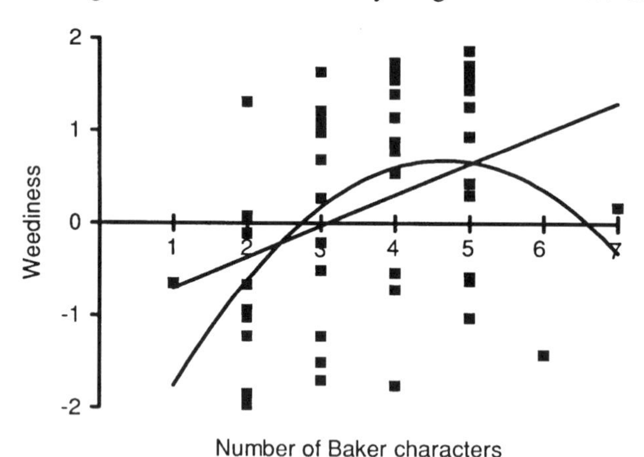

Figure 3.2 Relation of Baker characters to weeds. From Williamson (1993), with the addition of the linear and quadratic regression lines of weediness against the number of Baker characters. The equations of the lines are:
$\text{ws} = -1.03 + 0.333\ \text{bs}$, ($P = 0.013$, $r^2 = 0.125$) and
$\text{ws} = -3.273 + 1.689\ \text{bs} - 0.181\ \text{bs}^2$ ($P = 0.004$, $R^2 = 0.214$),
where ws is weed score and bs is Baker character score.

Baker score for each species. Those shown there are for 49 British annual plants. The Baker score was found by devising objective criteria derived largely from the Ecological Flora Database (Fitter and Peat, 1994). The weediness score is the combined opinion of 65 scientists (Perrins, Williamson and Fitter, 1992a,b). Remarkably, not one scientist agreed exactly with another; there is a large subjective element in declaring a plant at least a minor weed. Although a linear regression is significant ($r^2 = 0.125$, $P = 0.013$), the curvilinear one is almost twice as effective ($R^2 = 0.214$, $P = 0.004$). Neither regression is of any value for prediction. It is pointless to use Baker characters in risk assessments.

The mistaken idea that weeds differ from other plants in many characters and that those characters are Baker characters was unfortunately endorsed by a US National Academy of Sciences committee (National Academy of Sciences, 1987). A plant can become a weed with no change of characters, just a change of habitat (Williamson, 1993). Such a change of habitat can be a natural accompaniment of an invasion (see section 3.4.5).

There is a similar mistaken view, endorsed in the same committee report, that a microbe has to acquire a suite of characters to become a pathogen. Again, no change is needed if the habitat changes. 'Although *Enterococcus faecalis* was once regarded as nonpathogenic, this opportunistic gram-positive coccus now ranks among the most troublesome hospital pathogens. It has intrinsic resistance to many antibiotics and a remarkable capacity for developing resistance to others' (Phillips, 1995). The habitat change here is that produced by the use of antibiotics in hospitals.

In sections 2.3.2 and 3.3.1 the success of biological control insects was treated as a possible, though evidently weak, comparison to invaders becoming pests. Goeden's (1983) scoring table for predicting such success was mentioned. The three characters picked out there as possibly more relevant to success than establishment are reminiscent of the sort of character Baker used. They were a long attack season, ease of culture and gregarious feeding. Myers (1987) looked more specifically at the causes of failure, the obverse of success. She discussed five factors that may be important: climate differences (see section 3.4.5), quality of the plant attacked, enemies of the control agent in the area of introduction (see section 2.5), genetic variability (see section 6.2) and abundance and enemies in the native habitat (see section 3.4.3). The quality of the plant attacked can be thought of as an unusual way of looking at sufficient resources (section 2.3.3). However, as with Baker characters, all these indicators are based on case histories of past introductions. Statistical tests of the predictive success of these indicators would be desirable, but are of course difficult.

3.3.3 Statistical studies of invader properties

There have been several quantitative studies of the properties of invasive species, and of weeds. They show both the sort of characters that may be important in allowing invasion success and the wide range of such characters (Williamson and Fitter, in press, b).

The invasiveness of cultivated pines *Pinus* can be predicted well from just three characters, mean seed mass, mean interval (in years) between large seed crops and minimum juvenile period (again in years) (Rejmánek, 1995; Rejmánek and Richardson, in press). Their discriminant function used the square root of the first two characters. Small seeds, short intervals and short juvenile period point to invasiveness. Remarkably, the function divides their 24 species of pines into two equal-sized, and very clearly distinct classes. When pines are invasive from plantations they are usually regarded as weeds, and perhaps this discriminant function is a weediness predictor rather than an establishment predictor. The characters are those associated in plants with ruderal plants. The function does much less well in predicting natural regeneration after introduction in 34 other pine species.

Scott and Panetta (1993) found a function that fitted well Australian agricultural weeds that had been imported from South Africa over 140 years ago. The best predictor was weed status in South Africa, but climatic range in South Africa and con-generic southern African weeds contributed. Perrins, Williamson and Fitter (1992b) found weediness related to early germination, a long flowering season and seed dispersal through animal digestive systems. Nevertheless, they found their discriminant functions poor at prediction.

With birds in Australia, Newsome and Noble (1986) found rather little of significance other than the propagule pressure already mentioned (section 2.4.2). This was a study of both foreign introductions and native translocations. Using both, there is an indication that nesting is relevant, but this probably results from native parrots being successful and foreign game-birds not (often from few propagules). More realistically, the suitability of the receptor habitat is important.

That was a major conclusion too of Griffith *et al.* (1989) on translocations of birds and mammals. Their conclusion on propagule pressure has been noted (section 2.4.2), but they found six other factors that entered their step-wise equation before propagule pressure. These were, in order: (1) threatened, endangered or sensitive species versus native game; (2) habitat of the release area, whether excellent, good or fair-to-poor; (3) release in core of historic area versus peripheral to or outside it; (4) programme length; (5) early versus late breeder; and (6) bird versus mammal.

Factors (2) and (3) relate to habitat, but are clearly more use with translocations than with introductions. The discussion on climatic matching (section 3.4.5) will show how difficult it is to recognize suitable habitats for introduced species. Factor (1) may relate to abundance, considered in section 3.4.3, and factor (5) to r (section 3.4.1). Factor (4) is perhaps propagule pressure in a different form, but may also have to do with the chance of favourable conditions at release, mentioned in section 2.5. Their equation should be a valuable tool for managers. It shows what could be done with more systematic information on invasions.

3.4 FACTORS THAT HAVE BEEN SUGGESTED AS PREDICTORS (CFP 4)

The only predictors of invasion success that appear repeatedly in the studies discussed so far are propagule pressure, suitability of the habitat and previous success in other invasions. Certain factors that seem obvious to some people, either from population dynamics or from general biology, may be difficult to measure, and so have not been used in statistical studies. Some may be related not so much to invasion success as to weediness or colonization. These include the intrinsic rate of increase, modes of reproduction and genetic structure, abundance and range in the native habitat and climatic matching that is the matching of the climate of the original area of the species to that in its introduced range. All these aspects will now be considered in turn, along with the possible effects of taxonomic isolation.

3.4.1 The intrinsic rate of natural increase

A population introduced into new suitable habitat will, as is well known, increase exponentially. Formally, we can write the differential equation

$$dn/dt = rn,$$

with the solution

$$n_t = n_0 e^{rt},$$

where n is the population size (n_0 initially, n_t at time t) and r the intrinsic rate of natural increase (see the box in Chapter 1). For an age-structured population, one in which birth rates and death rates vary with age, it is also well known that Sharpe and Lotka (1911), expanding on a 1907 note by Lotka, showed that the population would settle down to increase at a rate r when it had also settled down into a stable age distribution. In the initial stages, a structured population will increase at rates alternately above and below r (see e.g. Caswell, 1989), and the structure of the population will also change. The eventual behaviour is why r is called the intrinsic rate; it is a most important parameter in theoretical population ecology.

There is an unfortunate confusion in the literature because some people write r_{max} or r_m for what Lotka called r, and then use r to describe the actual rate of increase. It is then said that r (by which is now meant the observed rate of increase) will tend to zero as the population comes to equilibrium in numbers or density. This is contrary to Lotka's use, and contrary to the use of r in phrases such as 'r and K selection'. It is best to use r for the intrinsic rate, the rate at minimal density, and to use another symbol, such as dn/dt, for the actual rate of increase.

Another reason for avoiding r_{max} is that r is not necessarily, and indeed usually is not, the maximum rate of increase of which a species is capable. Rather, r is the maximum rate of increase under certain specified conditions. For insects in 'stored products', that is insect pests of flour, beans and other food stuffs, such as the beetle *Tribolium*, r is a function of temperature and humidity. It is also a function of the quality of the food. In general, r is a function of all relevant environmental variables, and is only at a maximum at one particular combination (Williamson, 1972).

Lotka was interested in, among other things, comparing different human populations, none of which was in a stable age distribution. His r is derived from tables of mortality with age, usually known as life tables, and tables of reproduction with age (the l_a and m_a functions described in the box in Chapter 1). That is, he subsumed the mortality that actually occurred, from diseases and parasites, from accidents and other causes, in the parameter r.

How this concept should be extended to non-human populations is not obvious. Crawley (1986) for instance, while excluding competition, natural enemies, mutualists and a 'mystery ingredient' (immigration) from r, nevertheless lists survivorship (and so by implication some forms of mortality) as a component of r. (Mutualists are species that help each other, by symbiosis or in other ways.) In practice, r is normally calculated from observations on laboratory populations. So mortality factors that act in laboratory culture are included while those that are only found in free populations are excluded. That approach is normal with microbes and arthropods. For species that cannot reasonably be kept as laboratory populations, which includes most vascular plants and vertebrates, the alternative is to estimate the mortality and fecundity rates at minimal density (ignoring Allee effects, adverse effects of low population density). This is difficult and has seldom been done, but some examples are given in section 4.3.4.

What has all this to do with invasion? It seems natural to many ecologists to assume that a population that reproduces faster than another will be a better invader. Like all assumptions in this field, it needs to be tested. Another reason for considering r is the popularity of the concepts r and K selection first put forward by MacArthur and Wilson (1967). K is the equilibrium that a population approaches with time, often called, somewhat misleadingly, the carrying capacity. Building on this idea, many people thought that K-selected species were more those of stable communities, particularly those where competition was important, while r-selected species were colonizing species, putting their resources into efficient reproduction, dispersing, perpetually finding new sites. Indeed, that is what MacArthur and Wilson themselves said. In effect they were saying that colonizing species, as defined in section 3.3.2, are r-selected. Colonizing species are not the same as invaders. So even if colonizers are r-selected, that is no reason to think invaders are.

All these generalizations should be treated sceptically. The concepts of r and K selection are not supported by surveys of data, nor do they fit newer sophisticated life history theory (Stearns, 1992). Generalization to an r–K continuum is equally unsatisfactory. A three-factor scheme has been put forward by many authors, and still finds favour. That it has been suggested several times independently indicates that it embodies some biological truth. It is certainly helpful for understanding the variety of life-history styles of herbaceous plants (Grime, 1987). But it is still controversial, and its usefulness in prediction or modelling has yet to be demonstrated (Williamson, 1989a).

The evidence that establishment depends on the intrinsic rate of increase is thin and contradictory (Williamson, 1989a). Lawton and Brown (1986) and Crawley (1986) tried to relate *r* respectively to invasion success and to success in biological control. Lawton and Brown said, 'the only certainty is that there are neither theoretical or empirical grounds for believing that *r* alone is the principal, or necessarily even an important, arbiter of invasion ability', but Crawley said, 'it is clear that for establishment, and for successful depletion of weed abundance, agents with higher intrinsic rates of increase are more likely to succeed'. In neither paper are there actual estimates of *r*; the conclusions come from measurements of components (Crawley) or correlates (Lawton and Brown) of *r*, which may partly explain the difference. Goeden (1983) also used components of *r*, namely fecundity and generations per year, for finding promising biological control agents (section 3.3.1). But the agricultural biological control agents of Goeden and Crawley are likely to be colonizers as well as invaders, while Lawton and Brown's species have more varied ecologies.

What is certain is that some successful invaders have rather low rates of increase. For instance, the fulmar (section 1.3.1) has a yearly increase of perhaps 6%. Most strikingly, the most successful invader of all, *Homo sapiens*, has a maximum rate of increase of only half that, 3% per year. Nevertheless, when trying to predict which of several related species is the most likely to be a successful invader, and particularly is most likely to become a pest, the rate of increase may be relevant. Rabbits *Oryctolagus cuniculus*, with a high *r*, are a frequent pest and hares *Lepus europaeus* with a lower one are usually not. But note that both have established themselves in Australia and New Zealand and that the hare as well as the rabbit is a pest in the latter (Daniel and Baker, 1986).

The main correlate of *r* used by Lawton and Brown (1986) was size. There is a well-known relationship across widely disparate organisms from viruses to mammals: *r* is negatively related to body mass (Fenchel, 1974). As Fenchel points out, most of the estimates of *r* come from colonizing species, and so may be higher than typical values for the living world as a whole. The relationship across phyla is

$$\log r = c_r - 0.26 \log W$$

where W is the mass and c_r a constant (Blueweiss *et al.*, 1978). However, within genera or families it seems to be quite common for the relationship to be positive or humped, with the highest *r* at the largest or intermediate sizes, or even just random. Williamson (1989a) gave some examples. Gaston (1988), following that up, found in insect orders an insignificant negative relationship for both length and weight with *r* in Coleoptera, insignificant positive relationships with weight in the Hemiptera and insignificant to marginally significant negative relationships with length in the same order. Stemberger and Gilbert

(1985) found a positive relationship of r to dry mass in eight species of rotifers in different families.

It would seem that the broad negative relationship of r to size only holds consistently at the level of orders or higher taxonomic ranks. At the levels at which comparisons should be made for predicting invasions, within genera preferably but within families or even within orders, the linear relationship is frequently null to positive. Middle-sized to large species are likely to have greater reproductive rates than their small congeners. Interestingly, this change of relationship with rank also applies to abundance in birds (Nee *et al.*, 1991). Positive and negative relationships are found at all taxonomic levels, with the balance from mainly positive (within genera) to mainly negative (within orders) switching at about the level of the tribe (a little used rank between genus and subfamily). Abundance can be equated to the K of r and K, and then across phyla a standard equation is

$$\log K = cK - 0.75 \log W$$

(Damuth, 1991) with W being mass and cK another constant. There has been much discussion of the scope of this equation (Blackburn and Lawton, 1994; Gregory and Blackburn, 1995) but that does not affect the argument here.

That both r and K should be negatively related to mass across distantly related taxa, but positively in nearly related ones may seem surprising to those who think of them as two independent parameters from the logistic equation (see the box in Chapter 7), often written as

$$\mathrm{d}n/\mathrm{d}t = rn((K - n)/K) \text{ or } \mathrm{d}n/\mathrm{d}t = rn(1 - n/K).$$

Mathematically, and when writing sets of logistic equations for several species at once, it would be more usual to write the equation as

$$\mathrm{d}n/\mathrm{d}t = rn - an^2$$

where r and a would be independent parameters, but then K is r/a and so related to r. On that interpretation, r and K might well be expected to vary in similar ways taxonomically. In reports of laboratory populations a frequent positive relationship between r and K seems apparent. Salisbury (1942) thought there was a positive relationship between seed output and commonness in closely related plants.

For now, it can only be said that the relationship of r, or for that matter R_0 (see the box in Chapter 1), to invasion success needs more critical examination, but may yet turn out to be important.

3.4.2 Reproductive and genetic characteristics

Related to the idea that the rate of increase is a particularly important character when analysing invasion, is the idea that invading species have definable genetic characters, such as being inbreeders or asexual, polyploid, heterozygous and so on. Barrett and Richardson (1986) and Gray (1986) have examined this question for invading species, Brown and Marshall (1981) for colonizing ones. They all find that successful invaders or successful colonizers may have any one of a large number of suites of characteristics. As Gray puts it, 'invading species have genetic characteristics which need only be sufficiently protean to ensure rapid expansion in the new environment'. He also notes that invaders into the British flora include only a few 'colonizing species'. Even among colonizing species, which might be expected to show consistent *r*-selected characters, Brown and Marshall point out that they are not a homogeneous group, but display a wide range of evolutionary pathways.

In fact, some successful invaders can be said to have no genetics at all. The first example in Chapter 1, *Veronica filiformis* and a boom-and-bust species *Elodea canadensis* (section 2.4.1), are both, at least in Britain, without sexual reproduction.

While these plant examples suggest that genetic variation is irrelevant to invasion success, Myers (1987) suggested some biological control failures came from lack of genetic variation, but her evidence for this was weak. Some invasions seem to have gone along with a progressive loss of alleles, presumably from founder effects. In the cynipid gall-wasp, *Andricus quercuscalicis*, there seems to have been a linear loss of genes with distance from Hungary to Scotland (Stone and Sunnucks, 1993), the total number of polymorphic alleles detected dropping from 20 to 10. This example is described more fully in section 6.4.2.

Evolutionary and other genetic effects are considered in Chapter 6. However, as far as predicting the success of invasion in general goes, genetics for the moment has little to offer, even though understanding genetic systems may well be critical in some individual cases.

3.4.3 Abundance and range in native habitat

There may be a relationship of invasion success to the abundance of a species in its native range, or the size of that range, or both. For range, Goeden (1983) gives 6 points in his scoring system (section 3.3.1) to 'cover the full range of the target weed'. Moulton and Pimm (1986) found a positive relation of range size and invasion success in passerines in Hawaii, while Roy, Navas and Sonié (1991) noted a positive correlation between the number of native climatic zones and the number of continents invaded in brome grasses. Daehler and Strong (1993) think range one of the best predictors of success in invasion.

For both range and abundance, Crawley (1987) found that 'insect species that are widespread in their native lands are significantly more likely to become established than species with local or patchy distributions'. Abundant and wide-ranging dung beetles are better invaders than rarer ones (Hanski and Camberfort, 1991). Ehrlich (1989) lists, without quantitative evidence, both 'large native range' and 'abundant in original range' as concomitants of successful invasion potential. Ehrlich was considering a wide variety of vertebrate invaders, and he lists 10 other concomitants, including large size, some features of ecological generalists, short generation times (high *r*?) and much genetic variability. This list is a useful indication of where statistical work is needed.

In general there is a weak positive relationship between abundance and range; the correlation coefficient can vary from zero to 0.8 (Hanski, Kouki and Halkka, 1993). McNaughton and Wolf (1970) were apparently the first to quantify this, and it has been much studied recently (Gibbons, Reid and Chapman, 1993; Lawton, 1993; Sutherland and Baillie, 1993; Gaston, 1994). A sampling effect contributes to this correlation (Hanski, Kouki and Halkka, 1993). Abundance may indicate population parameters that favour establishment. Wide range could indicate flexible or generalist species, or good dispersers, and so more likely to invade successfully. However, the evidence is rather against wide-ranging species being generalists or having wide niches (Hanski, Kouki and Halkka, 1993) and there is no good evidence that dispersal is important in establishment.

There are other ways of looking at the relationship. Rabinowitz (1981) distinguished three types of rarity: range, abundance and habitat specificity. She noted that the types were not independent in that certain combinations were seldom seen. But it is difficult, in practice, to distinguish between habitat specificity and habitat scarcity. Abundant species can be those that are specific to abundant habitats, and so probably have a wide range (Gaston, 1994). All in all, without more evidence, it seems simplest to think that abundance is the important indicator of a good invader, range simply a correlated variable. If abundance is equated with *K* (section 3.4.1), it could be said there is better evidence that *K* is important in invasions than that *r* is.

There is, however, evidence from biological control that abundance can, for good biological reasons, be a bad predictor. This is because of the role of specific enemies. A relatively rare species, rare because of such enemies, may perform excellently as a control agent once freed of those enemies. Thus Goeden (1983) gave 6 points, a high positive score, to potential weed control species that are 'subject to extensive mortality from specialized enemies including diseases, and relatively immune to non-specific enemies'. The next best, with 3 points, are 'subject to extensive mortality from competitors for the host plant, combined with a common occurrence'. The worst, with 0 points, have 'natural control largely effected by non-specific enemies or abiotic ecological factors'.

Myers (1987) considered that species that fail to effect control of weeds can be 'insects that are competitively superior among the native guild of insects attacking a particular plant species or those that suffer little natural mortality'. 'Competitively inferior species or those that have evolved with high natural mortality' are likely to be more successful as control agents. In a sense, it could be said that these entomologists are saying that potential abundance is important, actual abundance may be misleading.

3.4.4 Taxonomic isolation

A species that is, in some sense, ecologically distinct from any of those in the community being invaded, may be at an advantage. Some aspects of this are considered below in section 3.4.6 on vacant niches, others in Chapter 5 when the success of species such as cats, rats and goats are considered. The success of, say, introduced mammals in New Zealand is probably most simply treated as the consequence of vacant niches. But one aspect of this is that they are taxonomically remote from any of the native biota. Is it possible to find an element of taxonomic isolation which is distinguishable as an element in the success of invaders?

For instance, there are no native parrots (order Psittaciformes) in either Europe or Hawaii, yet the ring-necked (or rose-ringed) parakeet, *Psittacula krameri* has become established in both (Belgium, England, Germany and the Netherlands in Europe (Gibbons, Reid and Chapman, 1993), Kauai, Oahu, Maui and Hawaii in Hawaii (Hawaii Audubon Society, 1993)). However, although surprising and spectacular (and see section 3.4.5 below) can it be ascribed to taxonomic isolation? Perhaps the surprise is that no other parrot has established permanently in Europe, and probably only one other (the red-crowned amazon, *Amazona viridigenalis*, a Mexican species) in Hawaii, though parrots are the most successful bird invaders in Australia (Newsome and Noble, 1986). The Australian species are, with one exception, invaders from one part of Australia to another.

There seem nevertheless to be no quantitative studies of this point, the significance of taxonomic isolation, with any animals, even parrots. With plants, Williamson and Brown (1986), found an indication that species from small families were proportionately more likely to establish than those from large families. Of course, the absolute numbers invading from large families was greater. Taxonomic isolation is a possible factor in invasions that needs more investigation.

3.4.5 Climatic matching

All species have a restricted range, which means that they are found in certain climates but not in others. Do the limits of climate in the native range predict the limits of an invasion? Can a species invade an area which has a climate different from that it is used to?

When introductions are made deliberately, as for forestry or biological control, it is normal to select species or strains from a climatically matched place (Booth, 1985; Booth *et al.*, 1987). Goeden's (1983) table quantified this for biological control insects. He gave scores from −10 for 'ecoclimate of area of introduction considerably harsher than the entire native range of agent' to +6 for 'ecoclimate of area of introduction and native range are similar'.

DeBach (1965) gave a classical view of the importance of climatic matching with parasitoids of the Californian red scale insect, *Aonidiella aurantii*. He discussed four species of the hymenopteran parasite *Aphytis*, a chalcid, which have reached California. *A. chrysomphali*, from the Mediterranean, arrived by accident about 1900 and spread throughout and became common in Californian citrus groves by the 1920s. *A. lignanensis*, from South Asia, was introduced in 1948 and displaced *A. chrysomphali*, which was almost extinct in California by 1963. However, *A. lignanensis* was itself eliminated in the interior of California by *A. malinus*, from Pakistan and India, introduced in 1958, but *A. lignanensis* remained the dominant species in coastal areas. DeBach interpreted all this to indicate that climatically matched species do best. All these three *Aphytis* species taken individually could invade California, but other *Aphytis*, such as *A. fisheri*, from Burma, failed. More comprehensive information on *Aphytis* successes and failures throughout the world was given by Simberloff (1986).

Another fairly important indicator of the importance of climatic matchings comes from beetles. Atkinson, Briffa and Coope (1987) showed how seasonal temperatures in the Pleistocene and Holocene can be plausibly reconstructed from the preserved beetle fauna. Their method depends on the analysis of mutual climatic ranges. Almost every species of beetle has a well-defined climatic range, a compact zone defined by maximum summer and minimum winter temperatures of the localities in which it is found naturally. The overlapping climatic range, the mutual climatic range, of two or more species is naturally more restricted than that of a single species. So, from a list of species found at a given site, it is possible to put reasonably restricted estimates on the summer, winter and mean annual temperatures. The method seems a most sensitive one, and the temperatures indicated are certainly consistent with those of other evidence. This was the method Coope (1986) used to predict which Carabid beetle species could invade Shetland, Faroe, Iceland and Greenland if they managed to disperse there (Table 2.7).

The beetles perhaps give the best positive evidence that sets of species are limited by climate. Negative supporting evidence comes from the failure of climatically mismatched species. The budgerigar or lovebird *Melopsittacus undulatus*, a small parrot well known as a cage-bird, comes originally from the interior of Australia. Sharrock (1976) reported a feral population on the Scilly Isles, a small archipelago to the south-west of England. By 1975 there were more than 100 budgerigars on the island of Tresco, and some on four other islands. In Lack's (1986) matching Atlas, they have 'softly and suddenly vanished away' (Carroll, 1876). In fact the last birds died in the winter of 1976/77 (Long, 1981). Similarly canaries, *Serinus canaria*, have often escaped, occasionally bred, but never established populations in Britain. Lockley (1947) tried to establish a canary population on Skokholm in 1939, and found that they were easy prey for migratory sparrowhawks, *Accipiter nisus*.

Nevertheless, there are plenty of exceptions to climatic matching. These are of two sorts. The first is where a species invades a climate new to it; the second where a species is successfully introduced into a climate new to it. The first is well known for some birds. Here are five examples.

The closest, allopatric, relative of the canary is the serin *Serinus serinus*, a bird of Mediterranean areas in the 18th century. It spread to central Europe in the 19th century, and by mid-20th century had spread north to the English Channel and North Sea, north-east to the Gulf of Bothnia and east to a line from there south to the Black Sea (Harrison, 1982). Except for some single pairs, the expected spread into Britain and Denmark has not happened (Dybbro, 1976; Gibbons, Reid and Chapman, 1993). At the same time two other passerines, the greenish warbler, *Phylloscopus trochiloides*, and the yellow-breasted bunting, *Emberiza aureola*, have spread from central Russia towards the Baltic (Timofeeff-Ressovsky, Vorontsov and Yablakov, 1977). The fulmar (section 1.3.1) and the collared dove (Figure 4.1 and section 6.2) are two other bird species whose spread seems unrelated to climatic factors. Possibly the spread north of the serin has been limited by climate.

In cases like these five bird species, while habitat change is a possible partial explanation, it is tempting to postulate a genetic change, even though no morphological change has been noted. A remarkable feature of Coope's beetles is how they have not changed morphologically over a million years, but there may nevertheless have been genetic changes affecting other characters. The mutual climatic range does not work on Miocene assemblages of beetles, even though living species are involved (Coope, 1987).

The second set of exceptions to climate matching are with deliberately introduced species. For instance, rabbits in the interior of Australia are found in much hotter and arid climates than those found in Europe. The Monterey pine, *Pinus radiata*, has a very small restricted range on the coast of California (Griffin and Critchfield, 1976), but is a most successful forest tree in many parts of the world, and an invasive species in several areas, including South Africa (Macdonald, Kruger and Ferrar, 1986; Richardson, Williams and Hobbs, 1994; see also section 3.3.3). The marine diatom *Biddulphia sinensis* came from warm, salty, tropical seas to establish itself in the cold, neritic seas off north-west Europe (see section 3.4.6). In Britain, the ring-necked parakeet, *Psittacula krameri*, 'the most incongruous of birds to find at large in the British countryside' is now well established in south-east England (Gibbons, Reid and Chapman, 1993; see section 3.4.4). Its native range is across Africa from Senegal to Somalia, and in Asia from Pakistan to Burma and Sri Lanka.

Examples of invaders showing both good and bad climatic matching are not difficult to find. Mack (in press) gives many plant examples. Wilson *et al.* (1992) compared the climate range of exotic plants in New Zealand with that in Europe. They found 'close agreement in climatic limits for some species (e.g. *Echium vulgare, Onopordum acanthium, Senecio jacobaea*), but considerable differences for others (e.g. *Verbascum virgatum*)'. All in all, climatic matching seems a fine example of a factor that ought to be of overriding importance and yet is on the whole a rather weak indicator or predictor.

3.4.6 Vacant niche and ancestral habitat

One model for the invasion of species into a community is that the successful invaders occupy a vacant or empty niche, while unsuccessful ones are excluded because their niches are already occupied. This model is at once helpful and confusing.

It is helpful because most ecologists find the concept niche helpful; it is confusing because different ecologists, and indeed sometimes the same ecologists on different occasions, use the term niche in different ways. Niche, like many terms in ecology, such as disturbance, stress, stability and competition, is useful in a preliminary, exploratory description, but becomes difficult to pin down in particular situations. So, the literature, and textbooks of ecology, are full of discussions of the different meanings. There are extensive discussions of history and usage of the term niche in Schoener (1989), Griesmer (1992) and Colwell (1992). Which of these meanings is relevant to models of invasion?

A distinction is often made between a habitat niche and a functional niche. The habitat niche is simply the range of habitats a species can or does occur in, sometimes distinguished by Hutchinson's (1957) terms of the fundamental and the realized niche. The functional niche 'is the place in the biotic environment' or 'the status of an organism in its community' (Elton, 1927), 'the position or role of a species within a given community' (Whittaker, Levin and Root, 1973). Even so, a species can function only in a habitat of some sort, and the limits of its habitat are the limits of its functioning; hence, the two concepts are sometimes difficult to distinguish. Still, as habitat is a reasonably well-defined concept in its own right, it makes sense, as has often been said, to restrict the term niche to mean the functional niche. This, after all, was the original usage of Grinnell (1917) and Elton (1927). Whittaker, Levin and Root (1973) put this succinctly, 'We suggest: 1. The term "niche" should apply exclusively to the intracommunity role of the species'.

Within the concept of a functional niche there are more restrictive concepts possible, in particular what Odum (1971) called 'the trophic niche', which he again derived from Elton (1927), referring particularly to an animal species, 'its relations to food and enemies'. This could be called the food-web niche, the position of a species in a food-web diagram. Another restriction of the functional niche is to the resource niche, bringing in the concepts of resource utilization curves and other tools of theoretical ecology. Resource niches as a concept partly overlap trophic niches, as both include the organisms fed upon or otherwise consumed. The trophic niche also includes enemies, the resource niche includes other resources such as nest sites, and both these are part of the functional niche.

It has been argued (Herbold and Moyle, 1986) that with the concept of resource niche, there is no such thing as an empty niche. Any new species coming into the community will take resources that previously went to another species. To put it another way, on this view the only empty niches are those which, when occupied, involve no interaction of any sort with other members of the community, with the implication, maybe, that the occupant of such a previously vacant niche is not a member of the community. Clearly, if you take the view that there are no empty niches, the invasion of communities cannot involve occupying empty niches.

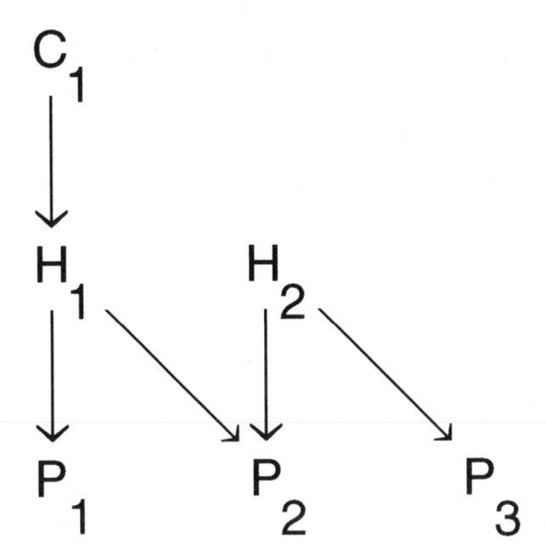

Figure 3.3 A diagram to illustrate the concept of a trophic niche. C, carnivore; H, herbivore; P, plant or primary producer. (See the text for an explanation.)

That argument runs counter to the way in which many ecologists think about niches. The difference may be brought out by considering trophic niches. Figure 3.3 shows a simple food-web of three plant species, two herbivores, and a carnivore. Each species has a distinct trophic niche. The two herbivore niches overlap, because each feeds on plant species P_2. Similarly, the plant niches P_1 and P_2 overlap because each is eaten by herbivore H_1. Similarly, the niches of P_2 and P_3 overlap. Now introduce a second carnivore C_2. If it feeds only on H_1, then its trophic niche is the same as that of C_1. If it feeds on H_1 and H_2 its niche overlaps with C_1. If it feeds only on H_2, many ecologists would say it is in a new niche, and it fills a previously empty niche. However, some of the resources that previously ended in H_2 now go to C_2. Herbold and Moyle (1986) following Whittaker, Levin and Root (1973) therefore argue that C_2 has not occupied an empty niche. Many ecologists would disagree. It is to some extent a matter of the meaning you want to put on the word empty. The food-web position for a carnivore feeding exclusively on H_2 in Figure 3.3 is clearly vacant, containing nothing, to use one dictionary definition of empty. Using empty niche in this sense at least allows a discussion of whether successful invaders always, often, sometimes or never enter empty niches, and so it is in this sense that the term will be used here.

In summary, the way the phrase vacant or empty niche is often used, and the way in which it is used here, means that a species occupying a vacant niche plays an entirely new functional role in a community, not that it does not use any resources previously used by other species. In this usage, a niche can be partly empty (or partly full) as well as completely full or empty.

On this view, some invasions may be described as invasions of empty niches. There was no poxvirus known in rabbits before myxomatosis (section 1.3.2). There was no large aquatic vole in Europe before the introduction of muskrats (see below). It is a commonplace that introducing either plants or insects without their natural enemies is one cause of troublesome invasions, with the corollary that they should be controlled by introducing natural enemies, and these enemies should be specific, attacking the troublesome invader. Biological control has been discussed sufficiently in sections 2.3.2, 3.3.1 and 3.4.3, and there is an overview of it in section 5.2.1.

One major difficulty with using vacant niches in this way is that, in practice and with the variety of relationships each species has with others, it is very seldom that a niche can be clearly said to be vacant, in the way in which we would say a second carnivore feeding only on H_2 in Figure 3.3 was occupying a vacant niche. Take the muskrat, *Ondatra zibethicus*, whose spread is discussed in section 4.3.2. There was no large aquatic vole in Europe before muskrats were introduced, but the water vole *Arvicola terrestris*, only one-sixth to one-tenth the weight of *Ondatra*, was nevertheless there, with many features of its ecology similar. There is an overlap in food, and also in predators such as the fox *Vulpes vulpes* and the pike *Esox lucius*. The justification for saying that their niches are separate, for saying the muskrat invaded an empty niche, is that there are no reports of population density effects of one species on another. Similarly, rabbits had other diseases, particularly bacterial, protozoa and helminth ones (Thompson, 1994) before myxomatosis. The extent to which a niche is vacant is, in practice, a post-hoc explanation. Post-hoc explanations are neither intellectually satisfactory nor much use in prediction.

In an attempt to assess quantitatively the importance of niches in invasions, Wilson, Hubbard and Rapson (1988) examined plants growing on shingle. They used classification and ordination to study the community role of 24 species that had been introduced to New Zealand from Britain. While the plants could be said, very broadly, to be in the same niche at Dungeness in England and at Upper Clutha in New Zealand, that only holds if all the plants are regarded as being in the same niche. The details of the relationships were remarkably different in the two places. 'The present results suggest that exotic species in New Zealand are occupying different niches from those of the native range'. At the point at which the niche concept might be useful in understanding invasions, in the detailed biology of individual species, it was found to be of no help.

Another example of a plant behaving surprisingly is *Verbascum thaspus*, great or common mullein (or just mullein). It is a biennial and a minor weed in temperate Eurasia, found on 'waste and rough ground, banks and grassy places, mostly on sandy or chalky soils' (Stace, 1991). Not surprisingly, it is an invader in Canada, listed as a weed (Crompton *et al*., 1988) 'in fields and on rocky or gravelly banks; common in British Columbia and eastern Canada, rare in the Prairie provinces'. The surprise is that it is a serious problem high up on tropical mountains. It has invaded the island of Réunion in the Indian Ocean (Juvik and Juvik, 1992). In Hawaii, where it has been since early in the 20th century it is abundant, forming continuous stands along roads, in leeward upland, 1200–3000 m, particularly on open ash and cinder substrates. It seems to be filling a niche similar to that of the endemic and fascinating silverswords, *Argyroxiphium* (Asteraceae or Compositae), which unlike mullein, are suffering from grazing by feral ungulates. Mullein in Hawaii also shows gigantism, fasciated stems, woodiness and becomes perennial, all unexpected and all possibly with a genetic basis (Juvik and Juvik, 1992).

Two more examples will show how what might be thought empty niches are, at best, only partly empty. In these examples, the niche was not fully occupied, but only somewhat occupied, before the invader came. This sort of situation may not be uncommon. One example is a marine diatom, the other a sub-tropical tree.

(a) *Biddulphia sinensis*

Biddulphia sinensis (Figure 3.4) is a planktonic diatom which invaded the North Sea in 1903 (Ostenfeld, 1908). It was probably brought by a ship and was a well-known species before the invasion, having first been described from Hong Kong harbour. It appeared to be a species of tropical waters, found from the Pacific around to the Red Sea in waters that are both warmer and of higher salinity than any of those around north-west Europe. As there are already many European species of *Biddulphia*, both planktonic and benthic, *B. sinensis* would seem to be an unlikely invader. Nevertheless, it

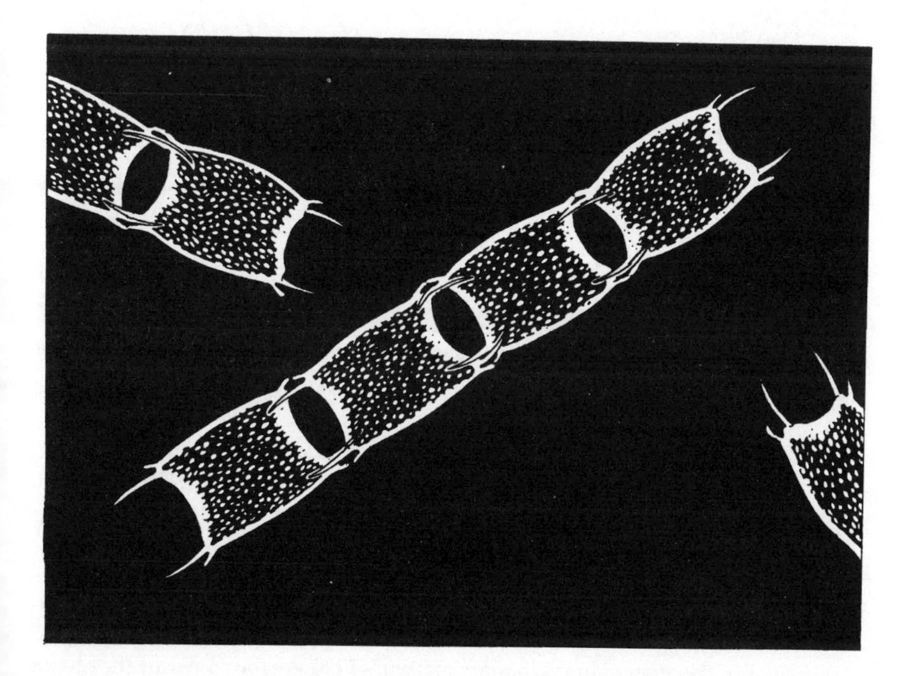

Figure 3.4 Short chains of the single cells of the diatom *Biddulphia sinensis*. Four cells occupy about 1 mm. (Illustration by Mike Hill.)

spread rapidly through European waters and established itself as a permanent member of the plankton in the least saline and, in winter, coldest areas of the continental shelf (Robinson, 1961). The reason for its success appeared to be that it had found an empty niche. 'It differs from all other diatoms found in the Irish sea in having its average peak of abundance in November' (Johnstone, Scott and Chadwick, 1924).

That this is only partly true can be seen in Figure 3.5. *B. sinensis* does have its main peak in November. It is, even so, not the dominant diatom in November, and the period when it is reasonably abundant, from September to May, covers all the colder half of the year, when other diatoms, including *B. mobiliensis*, are more common.

The invasion of *B. sinensis* in fact had a detectable effect on other plankton of Port Erin Bay, Isle of Man, the site of the data shown in Figure 3.5. A principal component analysis of the 22 most common forms, excluding *B. sinenis*, for each month of the 14-year period 1907–1920, shows this (Williamson, 1987). The first two components describe the average annual cycle; the third component reflects the invasion. Further analyses show that

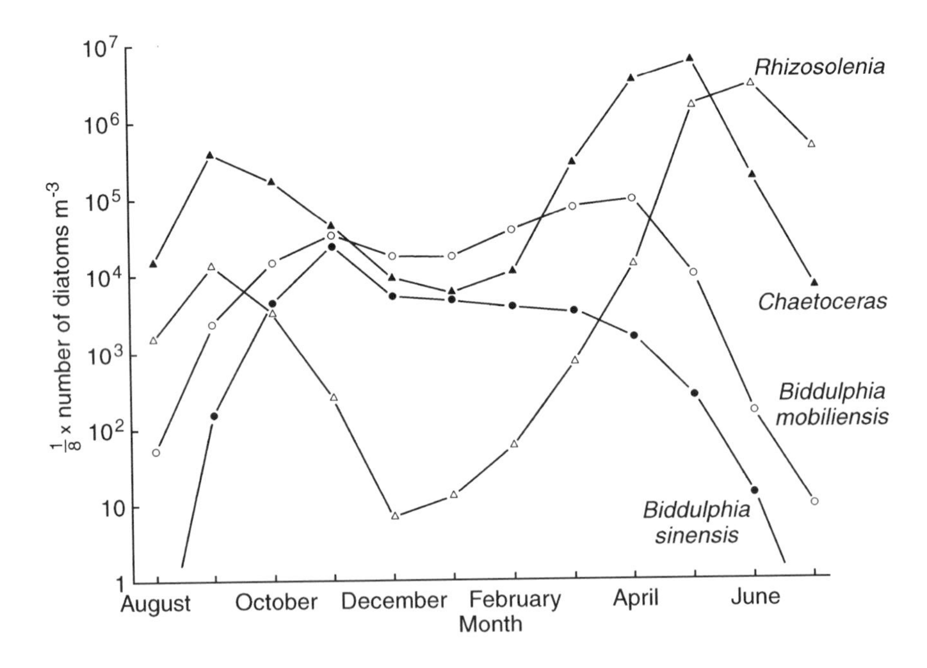

Figure 3.5 The average annual cycles of some diatoms (phytoplankton) in Port Erin Bay, Isle of Man, UK. The year is shown from August to July to emphasize the pattern in winter. *Biddulphia sinensis* invaded in November 1909; the data for it are for 1910–1920. For the other diatoms the data are 1907–1920. All the points are the squares of the means of the square roots of the monthly data in Johnstone, Scott and Chadwick (1924), and are plotted on a logarithmic scale. (From Williamson and Brown, 1986.)

the effect of *B. sinensis* lags a month or two behind its population size, and that the effect is most marked on other diatoms at particular times of the year. For instance, the population growth of *Rhizosolenia hebetata semispina*, a species that peaks in June, is depressed from January to March by high populations of *B. sinensis*. The summer peak of *R. h. semispina* is not affected. The apparently empty niche is only partly so.

(b) Melaleuca quinquenervia

The second example of a partially empty niche comes from Florida. *Melaleuca quinquenervia*, known simply as melaleuca, is an evergreen sub-tropical tree, expanding its range rapidly into natural ecosystems (Ewel, 1986). It is a myrtaceous tree from Australia. In its native habitat of coastal lowlands it forms open, nearly mono-specific stands that burn regularly. It is almost perfectly adapted to fire, burning fiercely peripherally but resprouting and germinating freely after fire. It can grow on most soils in south Florida, but is checked by frosts further north. It tolerates extensive flooding, moderate drought and some salinity.

South Florida has a mosaic of pine forests on the drier land, and Cypress forest in swamps, and the dominant species *Pinus elliottii*, slash pine, and *Taxodium* spp. respectively, thrive with occasional fires. In the transition from pine to the swamp edge, which is dominated by pond cypress, *T. ascendens* or *T. distichum* var. *nutans*, there is a zone, where the water table is at about ground level, characterized by stunted pines. Further out in the swamp, the dominant is the bald cypress, *T. distichum*. Melaleuca invades and dominates the ecotone between the pine and pond cypress (Figure 3.6), suppressing both the stunted pine and much of the pond cypress (Ewel, 1986).

Although melaleuca behaves as if there were a vacant niche, there were in fact trees, albeit poorly growing ones, in the niche before it arrived. The important factor is its ability to dominate in the niche. It seems simplest to regard this invasion as one determined by a suite of species specific characters.

At present, it looks as if models of invasion based on niches will be as disappointing as other community studies of niches. Niche analysis as an explanation of community studies is controversial with animals, and sometimes totally unsuccessful with plants (Mahdi *et al.*, 1989). Analysis of which factors react to population density rather than to attempt to study all factors as involved in the consideration of niches, might be more rewarding (Williamson 1957, 1989a; Law and Watkinson, 1989), but will certainly not be easy.

3.5 CONCLUSION

An invader can be any sort of species going into any sort of habitat. All systems are, apparently, invasible. Generalizations about invaders over too wide a taxonomic range, such as all species, or all insects, or all angiosperms, invariably have too many exceptions to be useful.

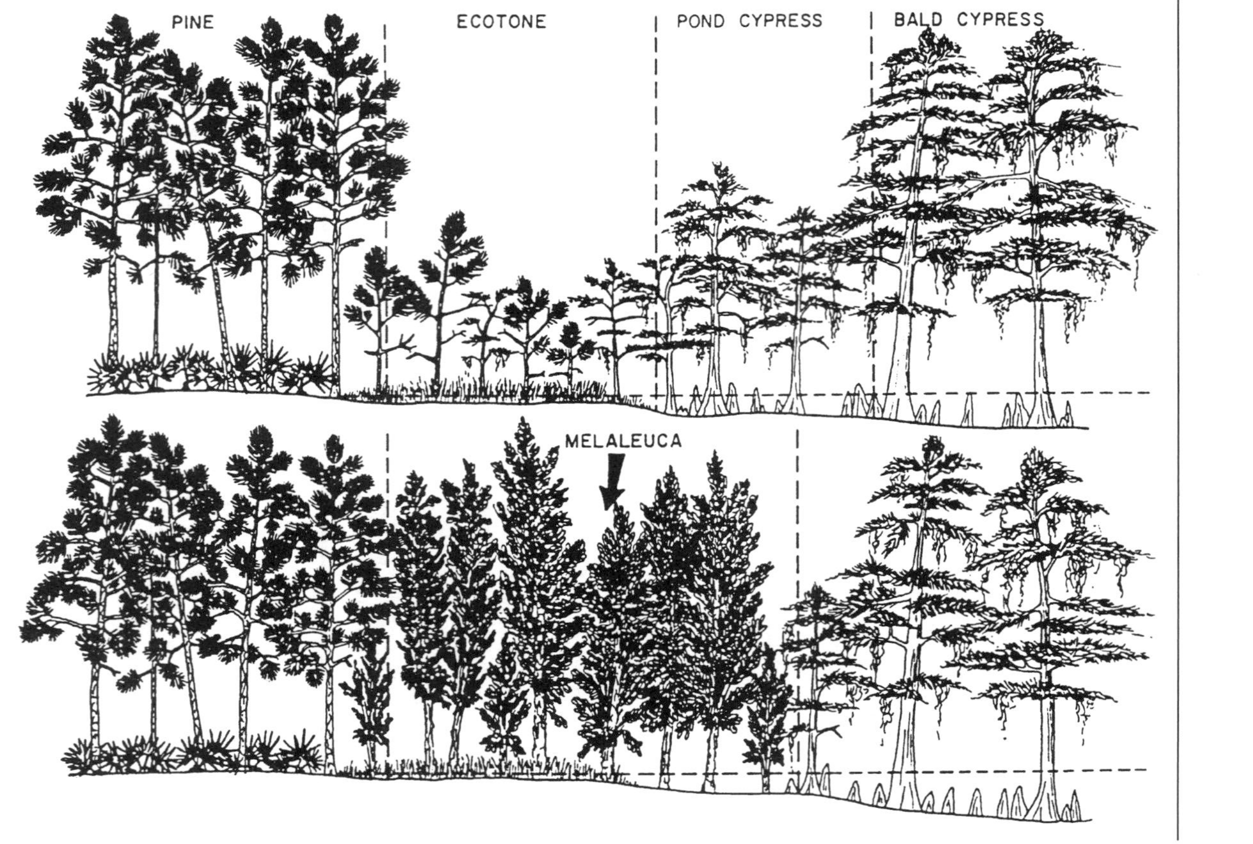

Figure 3.6 A diagram of the invasion of the ecotone between pine and cypress, in south Florida, USA, by melaleuca from Queensland, Australia. Note the nil effect on bald cypress, the small effect on slash pine and the major effect on pond cypress. (From Myers, 1984, with permission from

However, it is possible to generalize, for instance about pines, or translocated birds and mammals (section 3.3.3). The most robust generalizations seem to involve propagule pressure, habitat matching and previous success at invasion. Possibly propagule pressure, in the sense of reproductive rate after invasion, explains the rather varied results found in relation to the intrinsic rate of natural increase, and the weak relationships to range and abundance in the native habitat.

The role of taxonomic isolation and vacant, or partially vacant, niches will perhaps be clearer when there is a better understanding of community structuring. That topic is dealt with in Chapter 7 but, to anticipate what is said there, it is difficult as yet to do more than indicate what the difficulties are in understanding multi-species interactions.

Population dynamics is simpler and more effective when applied to the spread of species that have established, which I turn to next.

4 The process of spread

Conceptual framework point (from Table 1.1)

CFP 5 Spread can be at any speed in any direction; in analysed cases usually either as predicted by r (the intrinsic rate of natural increase) and D (the diffusion coefficient) or faster

4.1 INTRODUCTION

Broadly speaking, there are three phases to an invasion: (1) arrival and establishment; (2) spread; and (3) some sort of equilibrium. (These were discussed in Chapter 1, in relation to the conceptual framework of this book.) The previous two chapters were mostly about arrival and establishment, the ones after this one mostly about approximate equilibrium, what happens when an invader has spread as far as it can and its population is no longer increasing every generation. The middle phase, the subject of this chapter, is the most spectacular, the most studied, the most modellable phase of an invasion, and certainly the easiest to draw maps of.

Maps of spreading invasions are commonplace in books on invasions, such as Elton's (1958) classic, or Hengeveld (1989). From one or a few sources, the population spreads out like a wave from a stone dropped on a lake. The maps of rabbit (Figure 1.2) and fulmar (Figure 1.1) are typical. Over large scales, the wave is diffracted by changes of habitat, an effect particularly obvious with the fulmar, where the advance is affected by the location of coasts with cliffs. Two more are shown in this chapter, to go with particular analyses: muskrat, *Ondatra zibethicus* (Figure 4.4) and sea otter, *Enhydra lutris* (Figure 4.7). Similar maps can be found for all sorts of organisms, not just birds and mammals. For instance, the Africanized bee *Apis mellifera* in the Americas (see Figure 6.2), the cane toad *Bufo marinus* in Australia (Sabath, Broughton and Easteal, 1981), epidemics of influenza (a virus) (Patterson, 1986) and the Black Death *Yersinia pestis* (a bacterium) (Murray, 1993) and many others. These maps are probably all to some extent misleading, in that they imply there is a sharp boundary separating occupied and unoccupied territory. To counteract that, consider Figure 4.1, which shows the distribution of the collared dove *Streptopelia decaocto* (Figure 6.1), in Europe in 1948. At that time the species was spreading predominantly from the south-east to the

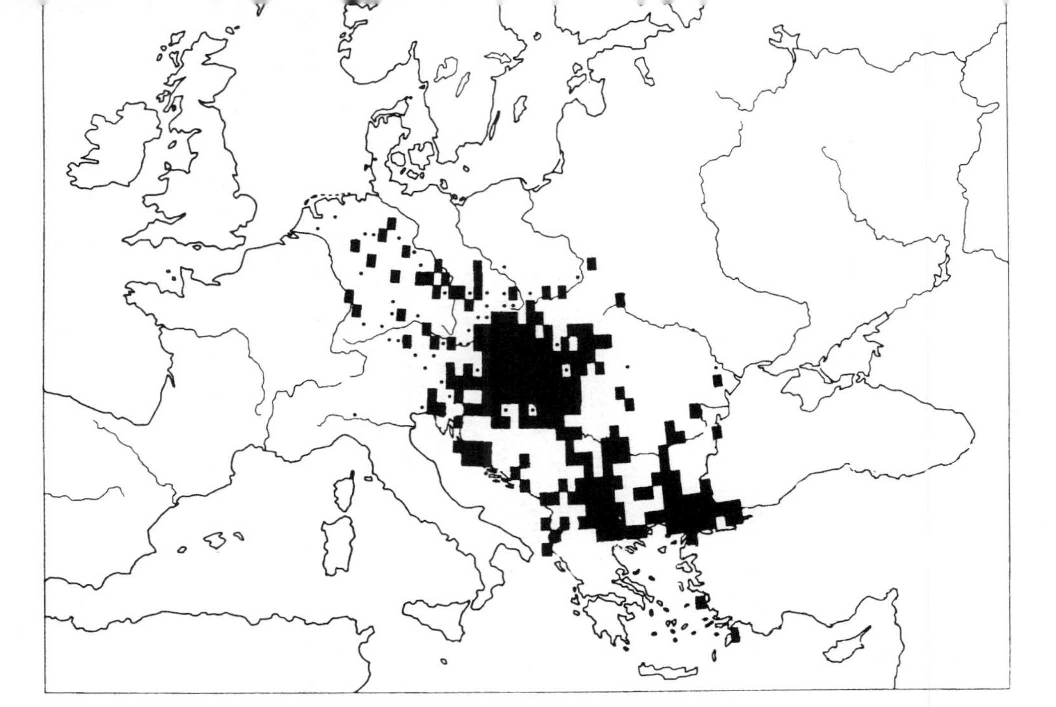

Figure 4.1 A map showing the state of the spread of the collared dove, *Streptopelia decaocto*, in 1948. This bird had a restricted distribution in the south-east Balkans in the early years of the century, and then spread, predominantly north-east. The fast spread started about 1930 and took the bird into northern France, the whole of the British Isles and southern Scandinavia by 1972. Note that at this snap-shot the advance in Germany is spotty, not as a block. The gaps in the Balkans may come from under-recording. Solid squares, probable or confirmed breeding; dots, isolated occurrences. (Modified from Glutz von Blotzheim, 1980, with permission from Aula Verlag.)

north-west as suggested by the pattern of the records. Figure 4.1 is no doubt an incomplete record of the bird at that time, but the spottiness of the spread is almost certainly real.

All these maps show two things. First, a roughly constant linear rate of expansion, i.e. the distance advanced in any one year is the same as in any other year during the expansion phase, and, second, variation with habitat. The variation with habitat is to be expected, but the constant rate of expansion needs a closer look. What are the exceptions? Why is it the rule, and are the exceptions predicted by variation in, or details of, that rule? What determines the rate? In this chapter, I will first deal with the mathematics of spread, in a simplified way, then look at some examples and tests of the theory. Then there is a discussion of a most spectacular spread, if somewhat unusual for an invasion, the spread of trees northwards in both Europe and North America after the last glaciation. Studies of pollen grains, and quaternary studies in general, have produced a wealth of information about this remarkably fast re-vegetation of areas that were tundra or under ice 15 000 years or less ago. Finally, I will return to theory with a discussion of spread with ecological interaction, looking forward to the ecological effects of successful invasions that will be discussed in the next chapter.

It could be said that spread is the least interesting phase of an invasion because it is just a constant ripple spreading over the landscape. But, as CFP 5 says (see the head of the chapter), the spread can be at any speed in any direction, and understanding those variations is interesting. If there is a need or a desire to control the invasion, then knowing how and how fast it will spread is useful. The faster the spread, the more difficult is control, the more important it is to act early. Seeing an invasion coming, as with the Africanized bee approaching the United States, allows and determines the time-span for counteracting or ameliorative actions. The theory of why the spread is at a constant rate involves some serious mathematics, which I will try to explain relatively simply. CFP 5 also notes that some spreads are faster than expected by theory. The reasons for that are still puzzling, and show there is still much to learn about even this superficially simple aspect of invasions.

4.2 THE MATHEMATICS OF SPREAD

A population of any organism newly established will usually do two things. It will increase in numbers and, as the initial population is small, normally it will soon do that at the intrinsic rate of natural increase, r. Second, it will disperse, it will spread out. (Sometimes, of course, it dies out.) The simplest description of dispersion is random diffusion. Random, here, does not require that the movements of an individual organism be random, but that statistically the set of movements in the population appear random. In practice, this happens in homogeneous environments, where individuals move equally in all directions, and where the movements of different individuals are fairly small and not correlated with each other (Holmes *et al.*, 1994). The conceptually simplest theory of spread deals with just these two processes, exponential increase and random diffusion. I will come back to complications later.

Putting those two factors together gives the deterministic linear partial differential equation

$$\partial n/\partial t = rn + D(\partial^2 n/\partial x^2 + \partial^2 n/\partial y^2) \qquad [4.1]$$

where n is the population size, t time and x and y are the spatial dimensions. The two parameters are r, the intrinsic rate of natural increase, and D the diffusion coefficient. An ordinary differential equation has just one independent variable, a partial one more than one. Here there are three independent variables, t, x and y. The equation is linear because n and all its derivatives come in linearly. It is deterministic, not stochastic, because there is no random variation in the values. Skellam (1951) introduced this equation to biologists.

A famous non-linear variant is the Fisher (1937) equation for the wave of advance of an advantageous gene. The same form results from substituting logistic growth (see section 3.4.1) for exponential growth in equation [4.1]. Population size under logistic growth increases from nearly 0 to K. Equation [4.1] with logistic growth is

$$\partial n/\partial t = rn(1 - n/K) + D(\partial^2 n/\partial x^2 + \partial^2 n/\partial y^2) \qquad [4.2]$$

Collectively, this pair of equations is sometimes called the Fisher–Skellam theory. Equations [4.1] and [4.2] can be described as reaction–diffusion equations, on which there is an enormous literature. Those wanting more mathematical detail might try Okubo (1980), Murray (1993) or Holmes *et al.* (1994) and the references they give. Here I will just state some results without trying to justify them.

As there is no preferred direction in equation [4.1], the population will spread out circularly. So, using the radial distance z (i.e. $z^2 = x^2 + y^2$; most mathematical accounts use r instead of z, but I want to use r for the intrinsic rate of increase throughout this book) the distribution of individuals at distance z and time t is

$$n(z,t) = (n_0/(4\pi Dt)) \bullet \exp(rt - z^2/(4Dt)) \qquad [4.3]$$

The e^{-z^2} term at the right means that the distribution has a tail like a normal (Gaussian) distribution, and there is a miniscule (and ignorable) probability of finding an individual at any distance out to infinity, so the distribution does not strictly have a place beyond which there are definitely no individuals. As there are just over 5×10^8 km^2 on the earth's surface, predictions of 10^{-8} individuals km^{-2}, or even a few orders of magnitude greater, can be taken as zero. There is no well-defined edge to the population, sometimes expressed as saying there is no front. This can be confusing, as the front of the wave (Figure 4.2) does exist. I will use edge to indicate the position of the foremost individual, front to mean the wave front.

To find the rate of advance, and to get over the problem that there is no mathematically definable edge, Skellam (1951) used the spread of the circle that excludes some threshold number, one or more of the population. Any number, provided it is not too big, can be used; the rate, surprisingly, is the same for all thresholds. The asymptotic form, the answer after some time, is the very simple equation

$$z/t = (4rD)^{1/2} = 2r^{1/2} D^{1/2} = 2\sqrt{(rD)} \qquad [4.4]$$

The rate of advance is constant and proportional to the square roots of the intrinsic rate of natural increase and of the diffusion coefficient. The more exact form of equation [4.4] can be written (Holmes *et al.* 1994)

$$z/t = (4rD)^{1/2} - b(t^{-1} \bullet \log t) \qquad [4.5]$$

where b is a constant derived from equation [4.3] and given in full by Andow *et al.* (1990, 1993) and Okubo (1988). The minus sign shows the initial spread is slower than the asymptotic speed. It is convenient to refer to all these equations [4.1–4.5] collectively as the Fisher–Skellam theory.

In some reaction–diffusion equations, the population advances as what mathematicians call a travelling wave, the population density being given by a constant-shaped curve moving at constant speed. The one-dimensional version of the Fisher equation is one such (Murray, 1993), and looks like Figure 4.2. In the two-dimensional forms of equations [4.1] and [4.2] the wave changes form slightly as it spreads, but is almost a travelling wave, looking like a normal distribution expanding sideways.

There are many possible elaborations of equation [4.1]. The population growth term rn can be any biologically reasonable function of n; the logistic of equation [4.2] is one such. The diffusion term can be replaced by functions that more accurately describe the movement of the population. Both can be modified to include age structure and stochastic variation. The telegraph model, with correlated dispersion, is discussed in section 4.3.4. Any biological process can be incorporated. The costs of doing so are greater mathematical difficulty and greater demands on the data when the theory is tested. This is

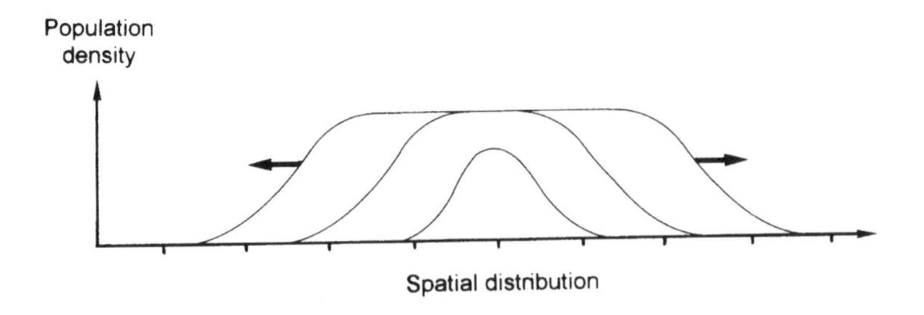

Figure 4.2 A diagram of a travelling wave. The lines show three successive fronts of the population as it spreads away from the centre. Note that, as there is an asymptote, the middle and outer fronts are the same shape. (Illustration by Mike Hill.)

an active field of research (Hastings, in press) which may well lead to new and unexpected results. Here I wish to do no more than give an indication of what is known.

In almost all cases, the rate of spread is determined by processes at the fringe of the population. If there is a travelling wave, these processes may be either at the front edge ('pull') or involve the whole front ('push') (Lewis and Kareiva, 1993), but in either case do not involve the population behind the front to any great extent. So, the population dynamics at low density are important. Often, as the rate of population growth is maximal at low density, $\partial n/\partial t = rn$ is a good description. Curiously, the apparent rate of growth at a point as the population passes by will often be about $2r$ (van den Bosch, Hengeveld and Metz, 1992, their equation 4.7 and Figure 7). The population moving in from higher densities more than compensates for that moving out to unoccupied areas. For some shapes of the travelling wave, the speed will be faster than that given by equation [4.5]. Andow *et al.* (1993) summarized some distributions from stochastic models suggested by Mollison (1977) that give speeds up to a little less than 30% faster than equation [4.4], so the effect can be expected to be small.

If the rate of population growth is not maximal at the lowest population densities, for reasons of age structure or of Allee effects, then the rate of advance will be slower. Allee effects are the name now given to a variety of low density effects: difficulties in finding food or a mate, greater risk of predation, inbreeding effects, and so on. Lewis and Kareiva (1993) gave the mathematics of this effect both analytically and with simulations. The asymptotic speed will be slower. How much slower depends on the mathematical function describing the Allee effect, but one-third to one-half of the speed given by equation [4.4] might be typical. The initial spread will be even more severely affected. Lewis and Kareiva suggested that the initial markedly slow rate of expansion that has been seen in some cases (see section 4.3 below) could be due to Allee effects. If such effects are present, the population may fail to spread at all if the initial invasion has too few individuals or occupies too little space, both of which are possibly factors in the importance of propagule pressure.

Age structure, i.e. individuals of different ages having different characteristics, can have a marked effect on population dynamics. Individuals hardly ever reproduce and disperse at the same rate and in the same fashion at all ages. How these variations with age can be incorporated into models of spread, using reproductive and dispersal kernels, was shown by van den Bosch, Hengeveld and Metz (1992) and in a more mathematical paper (van den Bosch, Metz and Diekman, 1990). A kernel in everyday language is the inner part of a (botanical) nut so, mathematically, it is the term used for a function that is placed inside another function. Adding age-specific reproduction and dispersal produces considerable mathematical complexity. Useful approximations for a variety of situations were given by van den Bosch, Hengeveld and Metz (1992). The main conclusion is that the pattern of spread is still the same, a radial spread at constant speed. In their case studies of four species of bird and the muskrat, taking the age structure into account increases the expected rate of spread by something of the order of 50%, but up to three times in one case. The reason is partly that in some of the birds they studied juvenile dispersal is greater than adult dispersal, but I will return to this point in section 4.3. Age structure effects could also slow down the rate of spread.

These age-structured models are still linear deterministic models. The general effects of non-linearity and stochasticity were discussed by Mollison (1991). He looked at a number of complex models from the literature. Broadly speaking, non-linear deterministic and linear stochastic models predict much the same behaviour as the more tractable linear deterministic ones. When there are differences in the detailed results, these may be more because of the values taken for reproduction and dispersal than because of differences in the models. Non-linear stochastic models are much more difficult, but roughly speaking they generally predict a slower rate of spread, with a maximum given by the results from the linear deterministic case, but they can sometimes predict a 'great leap forward' (Mollison, 1977). How complex a model to use will be determined by the quality of the data. Information on the distribution of dispersal distances is often rather poor, and on the variation of reproductive rate not much better, which limits the gain in understanding that might be hoped for from using more complex models. It is a good general rule to use the simplest possible model that captures the main features of the situation and that gives a reasonable fit to the data.

The critical feature in most models is the shape of the front of the advancing population. For the standard Fisher–Skellam theory it is, as noted above, the shape of the tail of the normal distribution, $\exp(-x^2)$. In almost all cases in the literature the shape is exponential, $\exp(-x) = e^{-x}$, or thinner, like the normal distribution.

Shaw (1994, 1995) has suggested that it could be thicker, and that that would have striking effects. He uses the Cauchy distribution, a standard mathematical distribution that looks quite like an exponential but has an appreciably thicker tail. Shaw argues that it might be appropriate in certain cases of wind dispersal. The mathematical effects are dramatic. The distribution has an infinite variance, and applied to dispersal means that the population shows no characteristic scale, and there is no wave front. New colonies appear any distance in front of the previous population. Shaw's simulations do resemble the patterns seen in the spread of plant pathogens. How realistic this model is remains to be seen.

However, the major prediction from most models of spread is that the population will enlarge at a constant speed along all radii at a speed of about $2r^{1/2} D^{1/2}$. Faster or slower speeds may call for models with age structure, or Allee effects or even non-linear stochastic models. The initial spread will be slower under many models, much slower or even negative if there are strong Allee effects. What do the data show?

4.3 TESTS OF THE THEORY

4.3.1 The measurements

To test the theory, to test if equation [4.1] or some more complex equation is a reasonable approximation, at least three things have to be estimated. From equation [4.4] these are the rate of advance of the population, z/t, the intrinsic rate of natural increase, r and the diffusion coefficient D.

To estimate z, and noting that populations never advance in exact circles, it is simplest to use the square root of the area occupied at various times. Various averages of distances along different radii could be used. The square root of the area is equivalent to the root mean square radius, and so will be larger than the arithmetic, geometric or harmonic means of the radii (Andow *et al.*, 1993), but the difference is negligible for reasonably symmetric spreads. When the spread rate is notably different in different directions, it is best to use estimates in each direction and avoid averaging them (Andow *et al.*, 1993). When spread is prevented in some directions by geographical barriers then, obviously, it should only be measured in directions open to the population.

The intrinsic rate of increase, r, can be measured in two ways. It can be calculated from age-specific fecundity and mortality rates. van den Bosch, Hengeveld and Metz (1992) put their results in terms of R_0, the expected number of females produced by a female in her life time, which is also derived from the age specific rates (see the box, Chapter 1). The other way of calculating r is to measure the rate of increase of the population at minimal densities from $d \log n/dt$. As noted above, this is not the same as the rate of increase at a point under the wave front.

The diffusion coefficient, D, is much the hardest of the three to measure. Three methods have been used. First, some authors, including me, have calculated what it would be if the theory was correct, using estimates of z and r, and then judging if the value is reasonable. Second, others have used the results of mark–recapture experiments. Third, it is possible to estimate D from the increase in the variance of the position of individuals in the population (Lubina and Levin, 1988), provided the whole population is involved in the spread, that there is no population behind the front that should be excluded. The third method can only be used reliably in rather restricted circumstances, and the first is not an independent measure of D. But the second may be unsatisfactory too, as it will often measure short-term dispersal and may be misleading about dispersion over a full year or a full generation.

Surveys of dispersal rates show that there are wide variations, depending on circumstances and the design of the experiments. In *Drosophila pseudoobscura*, a much studied western American fruit-fly, Slatkin (1985) concludes there is a large stochastic component of gene dispersal, because direct measurements give values of a few hundred metres while indirect methods gave much higher values. Similarly in plants, measured distances for seed dispersal are usually in metres, yet observations up to hundreds of kilometres are not uncommon for wind and bird dispersal (Hughes *et al.*, 1994).

Both r and D will vary with habitat and environment, with geography and climate. Usually there is only a single estimate for either parameter, though Andow *et al.* (1990) give a range of values. No one has derived a functional relationship for either parameter against other variables. So when the rate of spread is markedly different in different places, we are back to judging whether the observed variation is reasonably in accord with what would be expected. Indeed, it is seldom that any standard error can be attached to r or D, and without that it is difficult to judge whether the differences in fit of different models is significant. I will look at two well-worked cases, the muskrat and sea otter, in detail, including a lot of background biology, before giving a more general overview of a variety of studies.

4.3.2 Muskrat

The muskrat, *Ondatra zibethicus* (often misnamed *O. zibethica*), is a large vole adapted to watery habitats (Figure 4.3). It can live in freshwater (and also salt water) marshes, ponds, lakes, streams and rivers. It is found native throughout much of North America (Hall, 1981). It has a valuable pelt, often known as musquash, and this species was, with another rodent, the beaver, *Castor canadensis*, the mainstay of the North American fur trade in the 19th century. Because of its valuable fur, it has been bred widely in fur farms. As it is a digging animal that can bite through wire netting easily, it is not surprising that it has often escaped. Feral populations now exist from France, through Central Europe, Russia, Siberia, to China and on the island of

Figure 4.3 A muskrat, *Ondatra zibethicus*. Muskrats have a total length (nose tip to tail tip) of about 60 cm. (From Booth, 1968, see acknowledgements, p. x.)

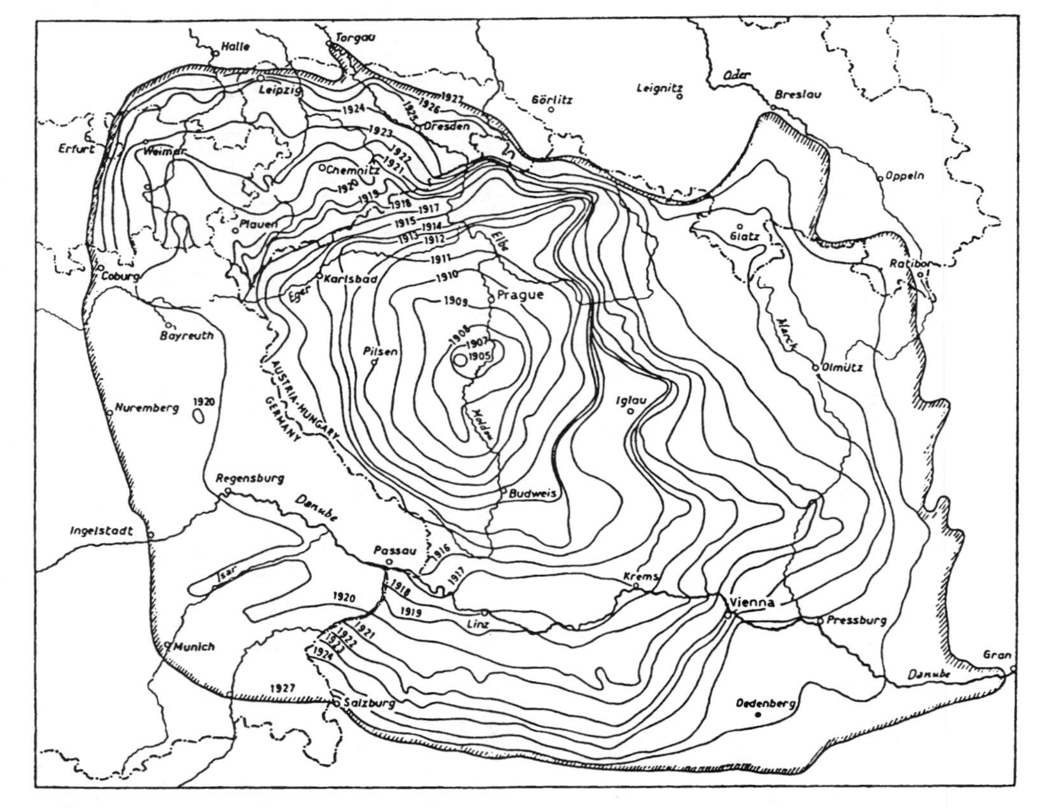

Figure 4.4 The spread of the muskrat in central Europe. This is a copy of the map in Elton (1958) which was adapted from the much larger, coloured, map in Ulbrich (1930). (Reproduced with permission from Mrs Elton.)

Honshu in Japan. There used to be feral populations in both Britain and Ireland.

The spread of the muskrat in central Europe has become the standard example of spread. Elton (1958) discussed it, and his map of the spread up to 1927 (Figure 4.4) has more contours than any of his other maps and is a beautiful example of quasi-circular spread. Skellam (1951), in a paper as classical in its way as Elton's book, used simplified contours to show that the square root of the area occupied was a linear function of time. I will return to more recent analyses after some more biology.

Voles and lemmings constitute the sub-family Microtinae or Arvicolinae, with about 130 species, in the rat and mouse family Muridae, which has some 1160 known species (Corbet and Hill, 1991). Voles are mostly small vegetarian rodents, with a body length of about 10 cm and weighing something like 30 g. Muskrats are the giants of the sub-family, with a 35-cm body length and a weight of 1 kg or more. They feed mostly on plants, but also on shell-fish, such as freshwater mussels, and they nest either in burrows or in dens. Dens, or houses, are conical and built in shallow water from plant material. Feral muskrats can be very destructive, undermining banks, dams, riverside trees and so on. By diverting water and choking channels with their burrows and debris they create marshes. As Matthews (1952) said, 'A person releasing a pair of musk-rats does far more damage, and causes far more trouble and expense in clearing up the results of his action, than a person releasing a pair of man-eating tigers'.

As an introduced species, the muskrat is of interest for at least three reasons. It is unusual as a successful transplant from North America to Europe; there have been many more introductions the other way, an example of propagule pressure. Its spread has been analysed in considerable detail; that is why this species gets a section to itself in this chapter. It is an example of a species which has been controlled, indeed exterminated, in some areas where it has been introduced, but it seems uncontrollable elsewhere.

Muskrats were first introduced to Europe, at Dobris in Bohemia (now the Czech Republic) in 1905, and were released or escaped in Russia, Finland, France and the British Isles in the 1920s. Two books have been written (in German) largely about these feral populations (Ulbrich, 1930; Hoffman, 1958). Verkaik (1987) outlined the present problems in the Netherlands. Muskrats reproduce fast, survive and disperse effectively. Parameter estimates are given below.

Attempts are made to keep the muskrat population under control in the Netherlands by trapping, apparently to little effect. Verkaik (1987) states that 230 000 were trapped in 1985. The Netherlands is only about 35 000 km^2 and much of it is, of course, not suitable for muskrats. As Verkaik said, 'The muskrat provides us with a striking example of the potentials of a prolific species encountering an apparently empty niche'. The niche of the muskrat was discussed above in section 3.4.6. Irrespective of whether you regard it as invading an empty niche in Europe, from its abundance in America it would be reasonable to have predicted that it would thrive in equivalent parts of Europe.

In contrast to the situation on the continent of Europe, muskrats have been eliminated in both Britain and Ireland. In Britain, the first recorded escapes were in Perthshire in 1927 (Matthews, 1952) and, with other escapes, muskrats were wild over an area of 65 × 30 km in the Forth and Tay valleys in Scotland. Other feral colonies were established in the Severn Valley on the borders of England and Wales, and in the Arun Valley in Sussex and near Farnham in Surrey in south-east England. There was but one colony, in a circle of about 30 km diameter, in Ireland around Lough Derg on the River Shannon (Warwick, 1934). It is perhaps surprising that more colonies were not established. Muskrats were known to have been kept, admittedly often in very small numbers, at 88 places in the British Isles. They are known to have escaped from 19 places (including those giving rise to the colonies just mentioned) and isolated muskrats were found at seven other places (Warwick, 1934).

The history of the elimination of the muskrat in Britain is short. Licenses were required to keep them from March 1932. The import and keeping of muskrats was forbidden from April 1933. Official trappers were appointed from June 1932. Populations were clearly lower in the summer of 1933, but it was thought unlikely they could be eliminated (Warwick, 1934). In fact, all breeding animals had gone by the end of 1935; two isolated males died in 1937. Slightly fewer than 4000 kills were recorded in the official campaign (Warwick, 1940).

In Ireland only a single pair was imported, and they escaped in 1927. An official campaign against their descendants started in September 1933. By May 1934 the population was extinct, 487 animals having been caught (Garvey, 1935; Moffatt, 1938). For six seasons, that implies an increase of 2.5 times per year. In my view, that is appreciably less than the rate of increase in central Europe, discussed below, but the estimates there vary from 1.77 to 4.0 times as will be seen. The remarkable speed with which the Irish population was extinguished perhaps suggests that muskrats found Ireland less suitable than Britain as a breeding ground, a surprising conclusion given the wetness of the Irish landscape.

These campaigns had their costs for other wildlife. In Scotland not only were 945 muskrats killed, so too were 2305 water voles *Arvicola terrestris*, 2178 moorhens *Gallinula chloropus*, 1745 brown rats *Rattus norvegicus*, 101 ducks (various species of Anatidae), 57 weasels *Mustela nivalis*, 36 stoats *M. erminea*, 18 herons *Ardea cinerea*, as well as small numbers of other species (Munro, 1935; Matthews, 1952; Usher, 1986).

In a more recent campaign against another large aquatic rodent, the coypu, *Myocaster coypus*, such deaths were avoided. Between 1981 and 1986 about 34 000 coypus were caught alive using cage traps. Coypus were then shot and non-target species released unharmed. Only 12 were caught in 1987, none (but two found dead) in 1988, and one in 1989 (Gosling and Baker, 1991). The history of the coypu in Britain is given by Smith (1995). It seems that coypus, like muskrats, have been eliminated from Britain. In contrast, efforts to eliminate American mink *Mustela vison* (Birks and Dunstone, 1991) and grey squirrel *Sciurus carolinensis* (see section 4.5) have been unsuccessful nationally.

Why was control of the muskrat possible in the British Isles? One possible reason is that it was started relatively early, not as early as it might have been, but nevertheless before the population became very widespread. Another possible effect, not contradicting the first, is that the muskrat may not in fact like the oceanic climate of those islands. The rate of reproduction appears to have been less in Ireland than in central Europe. The highest densities of muskrats are found in parts of North America with markedly continental climates, particularly round the Great Lakes and in the Rocky Mountains (Boutin and Birkenholz, 1987). The muskrat is absent from parts of British Columbia, even though it has been introduced successfully into Vancouver Island and the Queen Charlotte Islands (Banfield, 1974). Conceivably, the history of the control, or lack of it, of the muskrat in parts of Europe demonstrates the importance of climatic matching (section 3.4.5). Although the muskrat is such a pest in Europe, it is not regarded as a pest anywhere in America.

There have been four attempts to compare muskrat data with theoretical predictions, Andow *et al.* (1990, 1993), Skellam (1951), van den Bosch, Hengeveld and Metz (1992) and Williamson and Brown (1986). As the last is my own, I will try and keep my own preferences in perspective. The two papers by Andow *et al.* are part of one study and their publication was much delayed; they are contemporaneous with the Williamson and Brown paper.

All these papers used the data in Figure 4.4; Andow *et al.* and van den Bosch *et al.* also used other European data. The first question is whether the radial expansion is constant. The answer is yes and no. For the spread up to 1927, Skellam (1951) showed that the square root of the area gave a good straight line when plotted against time. In Figure 4.5 it can be seen that the square root gives a very satisfactory straight line and that the untransformed area and the logarithm of the area are both markedly curved. From those plots the expansion is 11.3 km per year, and that is the best average for the spread in all directions from the 1905 introduction up to 1927. Williamson and Brown noted that the speed along some radii seemed much greater. Andow *et al.* quantified this and found values from 10.3 ± 0.4 to 25.4 ± 0.8, which, remembering the average, shows that in a few directions the speed was over twice as fast. van den Bosch *et al.* (1992) found a marked slowing down after 1930, probably caused by an energetic trapping campaign. Their figures are 10.9 km up to 1930 (probably not different from the earlier value of 11.3) and only 5.1 km per year from 1930 to 1960. In France and Finland the spread was also relatively slow, from 0.9 ± 0.3 to 6.7 ± 0.8, and both accelerations and decelerations were found (Andow *et al.*, 1990, 1993).

Are these estimates of the rate of spread consistent with estimates of r and D? Is Skellams's equation, equation [4.1], and its asymptotic result, equation

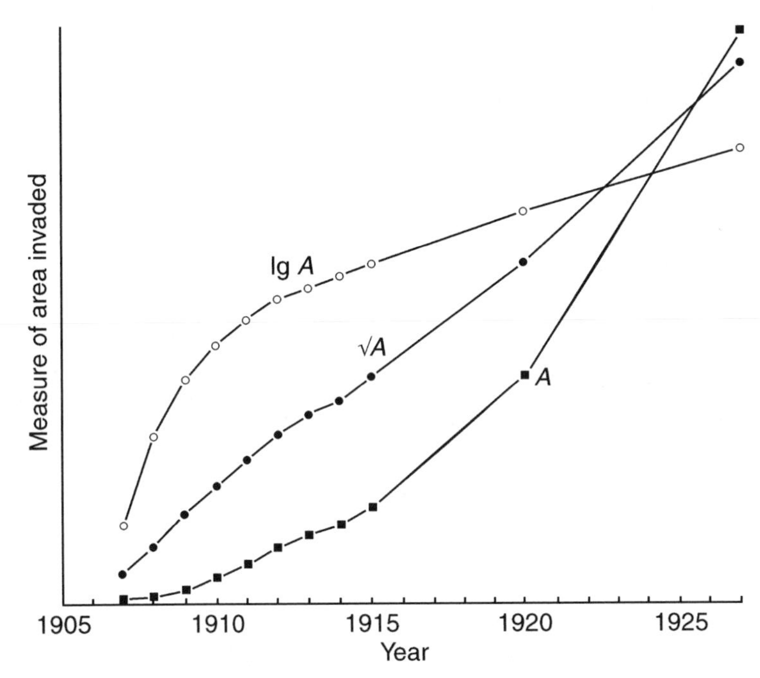

Figure 4.5 The spread of the muskrat, using data from Figure 4.4. The area (A), the square root of the area ($\sqrt{A}$) and the logarithm of the area (lg A) are each plotted against the year. The three scales on the ordinate have been adjusted to make the curves cross. (From Williamson and Brown, 1986.)

[4.4], a sufficient description, or is a more elaborate theory needed? At the moment it is difficult to be sure, as the estimates of r and D vary widely. The muskrat does disperse at all ages, so a major reason for using age-structured models, differential dispersal with age, does not apply. Andow *et al.* (1990) found mark–recapture estimates for D of 51.2 km^2 yr^{-1} from Netherlands data and 230.1 km^2 yr^{-1} from Finland. van den Bosch, Hengeveld and Metz (1992), taking a different study, estimated D as 10.125 km^2 yr^{-1} up to 1930 and 9.2 km^2 yr^{-1} from 1930–1960. Williamson and Brown, using the unsatisfactory back-calculation method, calculated a value of 22.97 km^2 yr^{-1} which, surprisingly, falls in the middle of all these estimates. There is an order of magnitude difference across these estimates, so it can only be said that the value of D is not known satisfactorily. It may well be markedly different in different habitats and different situations, and in any case the mark–recapture studies may not give the right answer for dispersion over the year.

Estimates of r are also very variable. It is perhaps easier to think in terms of λ the finite rate of increase, $r = \ln \lambda$. The Irish data above are the only estimate from the observed rate of increase and gave $\lambda = 2.5$ ($r = 0.916$). Using Leslie matrices constructed from the literature, Andow *et al.* found r from 0.2 to 1.1 (λ from 1.22 to 3), Williamson and Brown $r = 1.39$, $\lambda = 4$. van den Bosch *et al.* using field estimates for their life-table values found $r = 0.569$, $\lambda = 1.77$ for the earlier period and $r = 0.296$, $\lambda = 1.34$ later. With these differences, it is not surprising that van den Bosch *et al.* found their predicted rates of spread, 4.8 and 3.3 km yr^{-1} in the two periods, well below the observed, while the other studies thought the match reasonably good. Using their more elaborate age-structured model, van den Bosch *et al.* still only estimated 7.0 and 3.9 km yr^{-1}, 36% and 23% too low.

The low values in the report of van den Bosch *et al.* may have come from using data from established populations rather than from the initial low population density phase. It is notable that they have much higher death rates than Williamson and Brown (Andow *et al.* do not give details), and much lower reproduction rates. A major difference is that Williamson and Brown assumed two litters a year, van den Bosch *et al.* only one. There are no data to say which, if either, is right for newly introduced European populations, but it is known that Canadian muskrats can have up to three litters a year (Peterson, 1966), and ones further south up to six (Banfield, 1974).

Another species shows the difference in reproductive rates at low and at equilibrium density. For house sparrows, *Passer domesticus*, van den Bosch *et al.* note two studies 'in the density independent situation' giving 30 eggs a year, and use 12–18 in their calculations. Bird books suggest that perhaps around 12 is normal in established populations; for instance Summers-Smith (1988) quotes studies finding from 8.4 to 15.6 eggs per pair per year.

The muskrat shows that even in well-known, much-studied cases, there is still a great need for better field estimates. Well-established values of r and D at low densities and in different places are needed before critical tests of the predictions of different theories can be made.

4.3.3 Sea otter

The sea otter, *Enhydra lutris* (Figure 4.6) is a delightful animal to look at as it moves around in the sea over kelp beds along the Pacific coast of North America. So it is popular with tourists and naturalists. It eats sea urchins (which eat the kelp) and other fish and shell-fish, and it is not at all popular with fishermen. As a species confined to the coast, it provides a nice example of a one-dimensional spread. It is the World's largest mustelid, up to 1.5 m long and weighing up to 36 kg. (Mustelidae are the weasel family of carnivores.) Its original range was from Hokkaido (the northern main island of Japan), through the Kuril Islands, Kamchatka and the Aleutian Islands and down the western American seaboard almost to the southern tip of Baja California (VanBlaricom and Estes, 1988).

In the 19th century it was heavily hunted for its pelt and was near extinction when hunting was stopped by international treaty in 1911. It survives from the Kurils to Prince William Sound in Alaska, and has been successfully reintroduced at places along the Alaskan pan-handle and in British Columbia. In 1911 it was thought to be extinct further south. A small colony of about 50

Figure 4.6 A sea otter, *Enhydra lutris*, resting on the shore, others in the sea swimming on their backs. The total length (nose-tip to tail-tip) of a sea otter is up to about 1.5 m. (From Booth, 1968, see acknowledgements, p. x.)

was found in 1914 near Point Sur on the coast of the state of California, south of Monterey (Figure 4.7). The population started to spread about 1930, and the pattern of spread has been studied by Lubina and Levin (1988). Most of the coastline is rocky and suitable for the species, but there are extensive areas of sand and soft-bottom in the major bays (reached between 1972 and 1975 both north and south, and in 1981 in the south) which are unsuitable.

Allowing for the indentations and unsuitable habitat, it can be seen in Figure 4.7 that the spread has been at a fairly constant speed along the coast both north and south of Point Sur considered separately, but markedly faster in the south considering both together. Between 1938 and 1972 the rate averaged 1.4 km yr^{-1} in the north and 3.1 km yr^{-1} in the south.

Lubina and Levin (1988) estimated r from population counts both in California and Alaska (during the recovery phase). They give details of the reproductive biology, but do not attempt a life-table estimate. Their figure is 0.056 or 5.6% per year, which gives $\lambda = 1.0575$. D they estimate from the increase in the variance of the population location, giving 13.5 $km^2\ yr^{-1}$ to the north and 54.7 $km^2\ yr^{-1}$ to the south. Putting these in equation [4.4] gives satisfactory predictions of 1.4 and 3.8 km yr^{-1}. The difference between north and south remains unexplained, though Lubina and Levin offered some hypotheses that could be tested with more data. There is no obvious habitat difference, no biological explanation for the size of D, no reason to suppose r is different in the two areas. Water currents might conceivably be involved. The California current system is complex, but the dominant features are a southward current at the surface and a northward current along the bottom. The southward current has a maximum of around 25 cm sec^{-1} (about half a knot) at 15 to 20 km offshore (Huyer, 1983). Further, their attempt to estimate

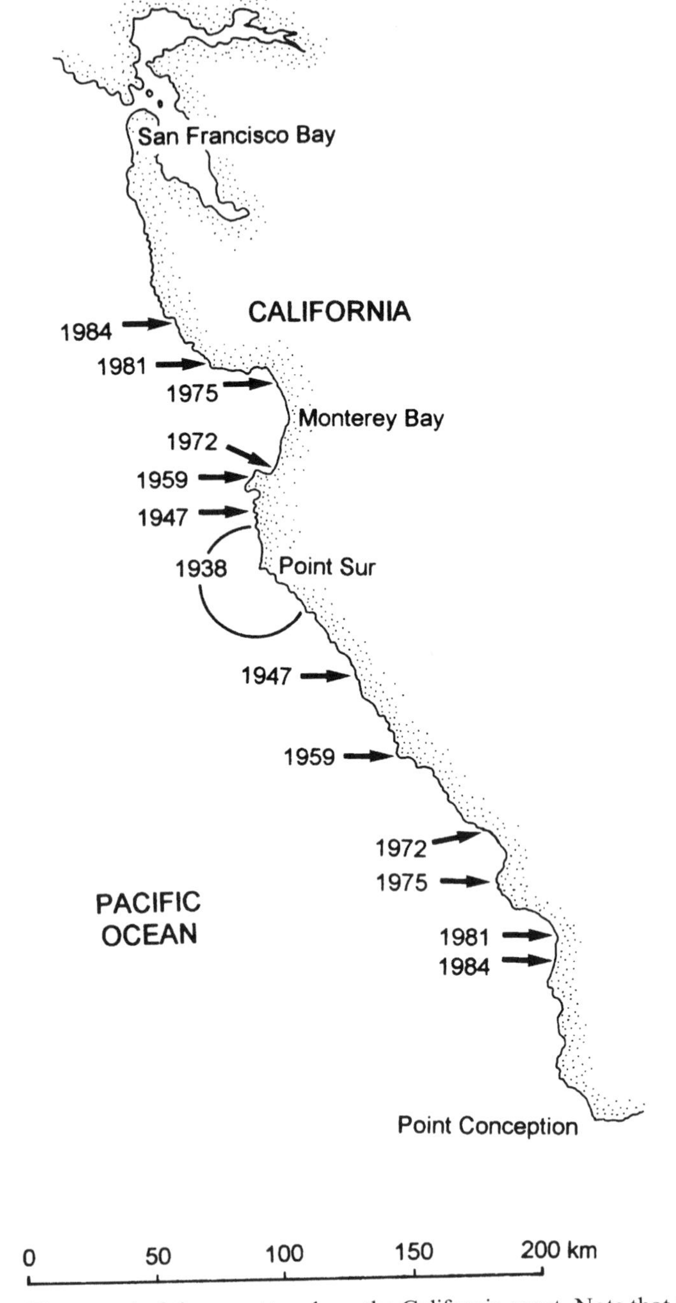

Figure 4.7 The spread of the sea otter along the California coast. Note that the years are at irregular intervals, and that the spread is generally, but not always, faster to the south. (Modified from Lubina and Levin, 1988, whose figure is Copyright University of Chicago Press.)

the total range expansion in one calculation is out by a factor of two. But in general, given the inevitable uncertainty in the estimates of *r* and *D*, this study perhaps is more in accord with the prediction of equation [4.1] than are the muskrat studies.

4.3.4 Other species

'By plotting radial spread of four species, we observed approximately linear rates of spread for muskrat and cereal leaf beetle [*Oulema melanopus*], accelerating rates of spread for rice water weevil [*Lissorhoptrus oryzophilus*] and small cabbage white butterfly [*Pieris rapae*], great leaps forward (Mollison, 1977) in small cabbage white butterfly, and pronounced geographical variation in all species.' (Andow *et al.*, 1993).

As that quote indicates, the standard Fisher–Skellam theory is a useful starting point, but is in no case completely satisfactory, and in several cases fails rather badly. A detailed discussion of case after case would be tedious. Instead, I will discuss the deviations from the standard theory that have been noted, and give examples. At the end of this section I will mention a set of tests of another model, the telegraph model.

Geographical variation in the rate of spread has been found whenever one species has been studied over a wide area. The quote above gives four cases; the collared dove (Figure 4.1) and the grey squirrel, *Sciurus carolinensis* (section 4.5 below) are others. The marked irregularities in many maps of widely spreading species show that geographical variation in rates must be the norm. But this is, of course, to be expected with the Fisher–Skellam theory, as both *r* and *D* can be expected to vary geographically. Locality-specific estimates of these two parameters are needed.

The other deviations from the Fisher–Skellam theory all concern the rate of spread and the pattern of dispersal. They can be divided into the initial rate, the shape of the sustained rate, the rate of the sustained rate, and types of dispersal.

The initial rate is expected to be slightly slower than the sustained, asymptotic rate under Fisher–Skellam theory. It sometimes seems much slower, though I know of no statistical test of this. A slow initial rate merges into an accelerating sustained rate. Graphically, this means that a plot of the square root of the area against time will be concave (like the lower line of Figure 4.5), and not straight. Okubo (1988) noted slow initial rates in three American bird invasions, house sparrow, starling *Sturnus vulgaris* and house finch, *Carpodacus mexicanus*. The effect lasted 5 to 15 years. The sea otter in California, as noted above, did not spread appreciably until about 1930, although hunting ceased in 1911, a delay of almost 20 years. Sea otter females are mature at 4–5 years old. The cereal leaf beetle in the quote at the top of this section had a lag of at least 4 years. The larch casebearer moth, *Coleophora laricella* had a slow rate of increase in the northern Rocky Mountains of perhaps 6 years (Long, 1977). A delay of around five generations is indicated for these cases. But such delays were not seen in any muskrat population, or in small white butterflies (called small cabbage white butterflies in America). Other cases have not to my knowledge been examined critically.

In those cases of lag, the sustained rate is reasonably linear. In other cases there is a longer acceleration. If the plot of the logarithm of the area against time is linear, then the square root plot is concave and there is an acceleration. This will be so when a logistic curve gives a good fit, as in the three species of *Impatiens* (Figure 1.5), up to the point of inflexion. Acceleration is also seen in the rice water weevil and small white butterfly in the quote at the start of this section and in at least one Finnish muskrat population and possibly some others. However, other Finnish populations showed a deceleration (Andow *et al.*, 1993). Some of the butterfly accelerations are so fast that they could indeed, and are in the quote, be called great leaps forward.

The rate of the sustained rate should be given by equation [4.4], but it is quite often faster than that. Instances include epidemics such as the black death (Murray, 1993), insects such as the cereal leaf beetle (two orders of magnitude too fast !) and the small white butterfly, birds such as house sparrow, starling, and cattle egret *Bubulcus ibis* (van den Bosch, Hengeveld and Metz, 1992) and mammals such as some muskrat populations (above) and the grey squirrel (below). The one terrestrial exception is the collared dove reported by van den Bosch, Hengeveld and Metz (1992); they found 43.7 km yr^{-1} for the spread north-west across Europe between 1930 and 1980, against estimates of 56.3 from Fisher–Skellam and 65.6 from their age-structured model. These estimates are about as much too high as their muskrat estimates are too low. The problem with evaluating all this is the lack of confidence in the estimates of r to some extent and D in most cases. Taking only estimates that are out by a factor of two or more eliminates the collared dove, starling and cattle egret, and leaves six examples of unexpectedly rapid spread. This result is incorporated into CFP 5 (at the chapter head).

A different problem with estimating D was found by Grosholz (in press). For ten marine examples from the literature he found that in all cases the predicted rate of spread to be higher than the observed. His analysis indicated that the problem was with the estimation of D rather than of r. Marine species had higher year-to-year differences in their rates of spread, as measured by coefficients of variation, than ten terrestrial ones. This was probably because of occasional long-distance dispersal events. Grosholz (in press) suggests that long-distance dispersal in the sea may often not lead to establishment. If so, there are problems in defining what is meant by D. His ten cases cover an ectoproct, *Membranipora membranacea*, three gastropods *Littorina littorea*, *Philine auriformis* and *Tritonia plebeia*, two bivalves *Mytilus galloprovincialis* and *Perna perna*, two crabs *Hemigraspus sanguineus* and *Carcinus maenas*, a barnacle *Elminius modestus* and a tunicate *Botrylloides diegensis*. The importance of larval dispersion for so many marine benthic organisms suggests that the methods of van den Bosch *et al.* should be tried.

The last deviations from theory, if such they can be named, are two cases in which there are two rates of dispersal. First, in the cholla cactus, *Opuntia imbricata*, in Texas there may be passive dispersal by seeds or stems falling off the plant or active dispersal by livestock and wildlife (Allen *et al.*, 1991). As they could not estimate *r*, their observations were limited to one generation. Second, Okubo (1988) suggested that 'jump-dispersal', perhaps another name for a great leap forward, was important at some stages of the spread of the house finch, but if so it only affected one of his eight data points. Such mixtures of dispersal styles are to be expected in many species, and certainly seem to be important for many of the forest trees discussed in section 4.4.

The final test of the theory is rather different. Holmes (1993), pointing to the biological unreality of the assumption of random diffusion in the Fisher–Skellam model, estimated parameters for another partial differential equation, the telegraph model. In this, the movements of individuals are limited and correlated. She compared the estimated spread in the two models for six species: (1) rabies, the virus of hydrophobia carried, in Europe, primarily by foxes *Vulpes vulpes*; (2) *Yersinia pestis*, the bacterium of black death; (3) *Pieris rapae*, the small white butterfly which she calls the cabbage butterfly (without the word small that name means *Pieris brassicae* to British lepidopterists); (4) *Lymantria dispar*, the gypsy moth which disperses as a first instar larva as the adult female is flightless; (5) *Streptopelia decaocto*, the collared (turtle-) dove; and (6) *Sturnus vulgaris*, the starling. To her evident surprise, the estimates from the two models generally agreed within 5%, those for rabies being almost identical. For the small white butterfly she derived four pairs of estimates which differed by 0.5, 0.6, 5.6 and 7.3%, the last two being the only differences over 5%. So, although the telegraph model could give estimates very different from the Fisher–Skellam ones, in practice it does not. Fisher–Skellam remains an acceptable standard model.

4.3.5 Conclusions on tests of the theory

The major difficulty arising from field data in constructing satisfactory models of spread is that the rate of spread seems often to be strongly influenced by local conditions. While it is obvious that *r* and *D* will vary with habitat, there are no good measures of this variation outside laboratories. It is difficult, shown most clearly perhaps with the muskrat, to get consistent estimates of either of these parameters. Without good estimates, tests of the comparative performance of different models are problematic.

In the studies above, the Skellam model, equation [4.1], often seems to give a reasonable fit, but also often seems to predict too slow an advance. But the species studied have, in general, been chosen because they show conspicuous, much discussed and certainly mappable spreads. Are they typical of invasive species as a whole? I can offer two indications that they are perhaps not. One compares two birds, the other looks at a set of species.

The spread of the house sparrow, *Passer domesticus*, was studied quantitatively by Okubo (1988) and van den Bosch, Hengeveld and Metz (1992), and the rate of spread was 15–30 km per year. The Eurasian tree sparrow, *Passer montanus*, only spread at about 1 km a year in the American mid-west. The invasion biology of these two species was compared in section 2.4.3. In established populations, Summers-Smith (1988) reports that tree sparrows produce about 5.8 young per pair per year, house sparrows 4.5, both with much variation and affected by many factors, and so probably not significantly different. For a 20-fold faster rate of spread, rD, from equation [4.4], needs to be 100 times greater. Apportioning that 100 equally to the two parameters means increasing r by a factor of 10 and so raising λ to its tenth power, an improbable increase; increasing D by a factor of 10 is conceivable. It seems better to assume that the spread of the tree sparrow is slower than would be predicted if its parameters were known. Summers-Smith (1988) says that the house sparrow limited the spread of the tree sparrow in America. Competition would certainly slow down the spread, as will be seen in section 4.5. The possible competition of house sparrows with house finches, *Carpodacus mexicanus*, is discussed in section 5.4, where it will be seen how hard it is to get satisfactory evidence. Cases like the tree sparrow have not yet caught the attention of modellers, but could be common. Meanwhile, I have phrased CFP 5 (see the top of the chapter) cautiously.

The set of data that also points to caution is the rather limited information we have on the spread of British introduced plants. As reported in Williamson (1993), Crawley counted 56 species as widely naturalized, Fitter 196 as fully naturalized, and another 348 as locally naturalized. Widely naturalized means geographically widespread; fully naturalized implies that populations can be expected to continue indefinitely; locally naturalized that the population is small and vulnerable, and likely to go extinct. The first two categories count as established (see Table 2.2), the third only as introduced.

The contrast between fully and widely naturalized perhaps tells us something about spread. Widely naturalized species, like the *Impatiens* spp. and *Veronica filiformis* discussed in Chapter 1, will mostly have spread rather fast. However, if an introduction spreads at 200 m per year, it will still have only gone 20 km in 100 years. That time is about the right order for the typical established British invader. As will be seen in section 4.4 below, 200 m per year is perhaps a typical estimate for the rate of spread of British plants after the last glaciation. A range of 40 km (from spreading 20 km in all directions) would certainly score as fully naturalized, but not as widely naturalized. At this rate of spread, the majority of established species should be, as they are, fully rather than widely naturalized. The latter have either spread remarkably fast, often helped by human transport, or are the result of multiple introductions, or both.

Is 200 m per year the distance that plants might be expected to spread by Skellam's theory? Taking λ at 10 ($r = 2.3$) for a typical plant, and noting that mean dispersal is $(8D/\pi)^{1/2}$ (Williamson and Brown, 1986), a spread of 200 m per year implies a mean dispersal of around 105 m per year. As dispersal distances are almost certainly skewed, the median dispersal, the dispersal of the typical seed, will be less than this, but this figure of 105 m still seems larger than would be expected in most species. To put it another way, on our present sadly inadequate knowledge of r and D in plants, the Skellam theory may only predict a spread of a few tens of metres per year for many species.

The counts of fully and widely established British plants could mean that while many species are spreading at around the predicted rate, there are about equally many spreading too fast or too slow. That's just a guess, but it can be expressed in numbers. If the 56 widely established are spreading significantly faster than 200 m per year, 'too fast', then about as many, 56 again for the sake of argument, may be spreading significantly slower than this rate, 'too slow'. That would leave 84 of the fully established as spreading at about the 200 m rate, 'about right'.

All this wild speculation does at least suggest the published terrestrial examples are biased, and that species spreading too slowly could be quite common. That reinforces my view that the Fisher–Skellam equations are the right place to start when studying spread. Nevertheless, fast-spreading species will have more impact, and may be critical in determining the structure of some communities, as we shall see in the next section.

4.4 THE POST-GLACIAL ADVANCE OF FORESTS

In his classical paper on spread, Skellam (1951) discussed both muskrats and oaks, *Quercus*. From the record of oaks in Britain since the last ice age, he estimated a spread of about 300 m per year. He took the time to first full seeding as 50 years, giving a spread of 15 km per generation. Even with Skellam's deliberately high figure of nine million acorns per oak, it is clear that the rate of spread requires something other than acorns rolling away from the tree.

Since then, a great deal has been learnt about the rate of the northwards advance of forests since the last glacial maximum. That maximum was about 18 000 years ago. At that time the northern limit of European forests was in the Mediterranean region, for North America near the Gulf of Mexico. By about 6000 years ago the ice caps had retreated to around their present positions and most species had reached their present limits. That is a very broad brush picture of a complex pattern.

The pattern of spread of different trees can be followed by pollen analysis combined with radio-carbon dating, and various other less useful observations. Both the main techniques have limitations. Pollen analysis mostly tells us about genera; it is not usually possible to distinguish species in the same genus by their pollen. The pollen of the trees most studied is spread by wind, and so some pollen will reach points well away from where the species is growing. How much pollen in a sample indicates that trees are growing nearby? That is a question that has been much disputed (MacDonald, 1993).

On a continental scale, the cover of samples is patchy. There are fairly large areas for which there are, as yet, no pollen records. Birks (1989) used 135 sites to map spread across the British Isles, but still had no samples from the English midlands, an area of about 200 km square.

A difficulty with radio-carbon dating is that there is not a simple linear relationship between radioactivity and time. Many dates are quoted as uncalibrated year before present, that is using the half-life of ^{14}C to estimate dates before AD 1950. The calibration curve, the curve of radioactivity against time wiggles because of variation in the rate of formation of ^{14}C (Pilcher, 1993). As the latest calibration tables were only published in 1986–1987, work published before about 1990 needs to be recalibrated when that affects the conclusions. In the worst case an uncalibrated year can correspond to several calibrated years. Uncalibrated 4500 bp (before present, apparently 2550 BC) could really be 3307, 3235, 3177, 3163, 3134, 3112 or 3110 BC (Chambers, 1993), a range of almost 200 years. But most wiggles are shorter and less convoluted. This difficulty makes only a small difference to the calculations below.

Here, I am not concerned with the details of which species was exactly where exactly when; my aim is to get some feeling for the rates of spread of trees in a roughly 10 000-year period. It is quite common to call such spread a migration (Chambers, 1993). In many cases one particular species in a genus will lead the migration in one area. Averaging over pollen sites and years minimizes most of the other uncertainties. Huntley and Birks (1983) used intervals of 500 years on contoured maps of pollen density in Europe, Delcourt and Delcourt (1987) 2000 years in eastern North America. Whatever mechanism is involved in migration, the average speed for such long periods and large areas must be about right, though the variation in speed from year-to-year, or even century-to-century, remains unknown.

Birks (1989) plotted the distributions of eight genera of deciduous angiosperm and one conifer in the British Isles and was able to calculate the area occupied by each at various times. He found that for most of the period of spread for each genus, the square root of area plotted against time gave a straight line (Figure 4.8). Naturally as a plant came to a physical limit, a coast or a climatic limit, its rate of spread slowed down. All the lines bend over at the top. It can be seen that most are covering 300 km in less than 1000 years, though *Fagus*, beech, is taking about twice as long as that. That gives an average of 300 m per year in the initial fast migration, or about 150 m per year for beech. Birks used equation [4.4] with estimates of r from Bennett (1983) to estimate D as between 1.7 and 9.1 $km^2\ yr^{-1}$ (median 3.9, *Fagus* not calculated). That is a mean dispersal distance of several kilometres a year, while the generation is measured in tens of years. Bennet estimated r at a point in Norfolk, England. As the species would have been spreading over that point, from the argument on page 85, his values would be about twice too large. Correspondingly, Birks' estimates of D would be half the size they should be. All this confirms Skellam's view that these spreads did not come from random diffusion. All the values are much too large for the Fisher–Skellam model, as noted in CFP 5. How the trees spread is discussed below.

In eastern North America, Delcourt and Delcourt (1987) measured the rate

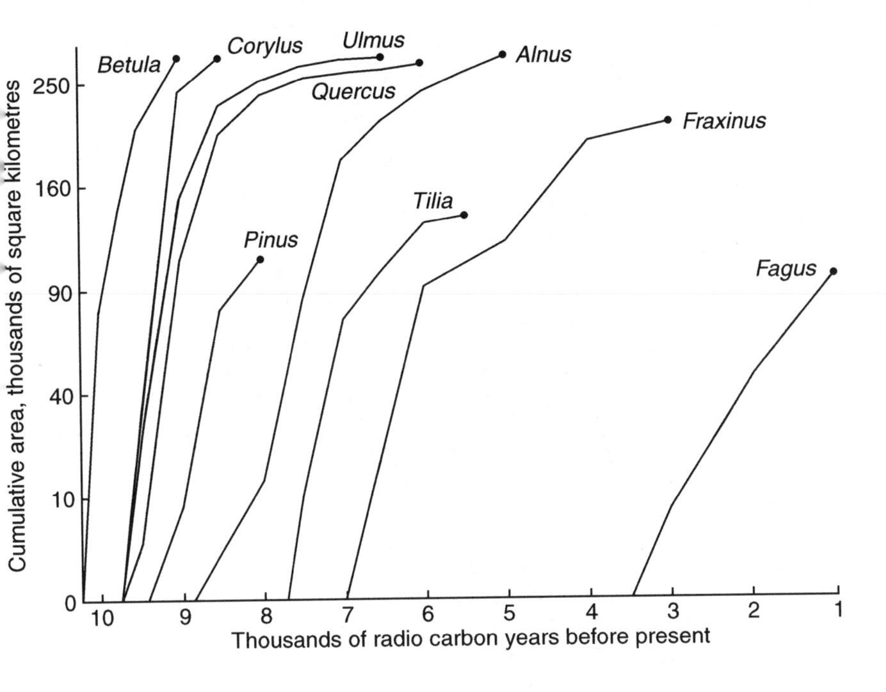

Figure 4.8 The areal spread of trees in the British Isles. The ordinate is the square root of the area, the abscissa time in thousands of years bp. (Modified from Birks, 1989, with permission from Blackwell Scientific Publishers.)

of migration of about 20 genera, some grouped together, and studied six in detail along five tracks (Figure 4.9). The tracks are an abstraction, each species advanced across a broad front, which differed somewhat from genus to genus. Each genus used only part of each track. Their tables give the rate for each genus on each track at 2000-year intervals, and, abstracting that, minimum, overall (or average) and maximum rates. The overall rate for each genus on each track is given in Table 4.1. The median rate is just over 150 m per year. Each genus behaves in its own variable way across the five tracks, so each track shows a different pattern for the six genera. Thus, 150 m per year for 2000 years is 300 km. There will be random variation in the estimates, but, even so, the differences between genera and tracks must be real, and require explanations in biological and environmental effects. That is not possible while the general phenomenon of fast migrations remains unexplained.

A principal component analysis of Table 4.1, using the rotation round the matrix mean, is presented in Figure 4.10. This variant allows genera and tracks to be plotted together. The first component separates the tracks from west to east and also the speed of the genera (modulated by their track pattern). *Populus* poplars and track five are the fastest. Component two contrasts tracks two and five, *Juglans* walnuts and *Picea* spruces. *Juglans* is twice as fast on

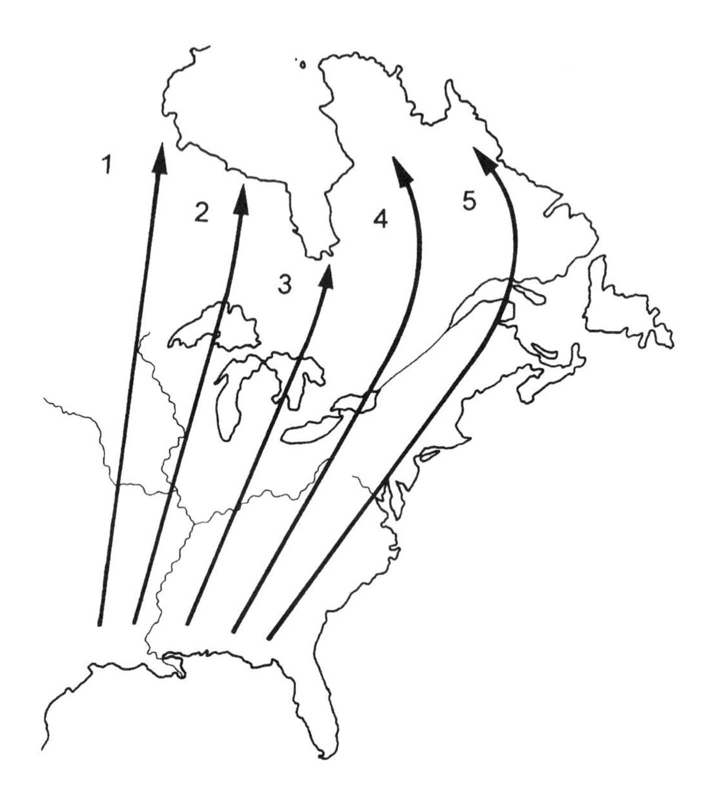

Figure 4.9 The tracks used by Delcourt and Delcourt (1987) when analysing the migration of trees in North America. Each species has travelled only part of any one track. (Illustration by Mike Hill.)

Table 4.1 Overall or average rates of spread of six trees over the five tracks indicated in Figure 4.9 in metres per year. The averages are over 10 000 years or more and, in some cases, include periods of zero or negative (backwards) spread. The symbol is the letter used for that genus in Figure 4.10.

	Genus (symbol)					
Track	*Carya (C)*	*Juglans (J)*	*Populus (P)*	*Quercus (Q)*	*Picea (S)*	*Larix (L)*
1	193	223	167	119	146	153
2	96	221	223	110	110	119
3	154	71	276	117	104	159
4	76	70	297	121	126	247
5	74	114	350	163	221	267

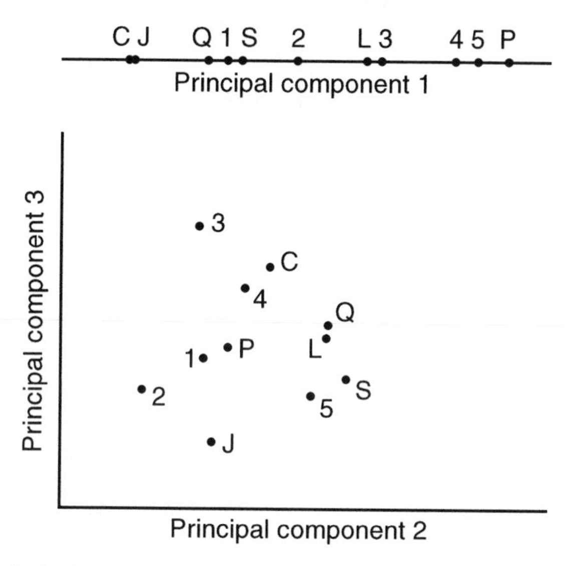

Figure 4.10 Principal component analysis of Table 4.1. First component shown on the top line, second and third components shown plotted against each other. 1, 2, 3, 4, 5 are the tracks shown in Figure 4.9. C, *Carya* (hickories); J, *Juglans* (walnuts); L, *Larix* (tamarack); P, *Populus* (poplars, cottonwoods, aspens); Q, *Quercus* (oaks); S, *Picea* (spruces).

track two as track five, *Picea* vice-versa. Component three contrasts track three against tracks two and five, *Carya* hickories against *Juglans*. But the important point is that each track and each genus is at a quite distinct place on the plot, each behaves differently from every other one. The two closest together are *Larix* tamarack (or larch, in other parts of the world) and *Quercus*, oak. *Larix*, a conifer, always has a much more northerly distribution at any one time.

It is possible to generalize about average and maximum rates, using Delcourt and Delcourt's data for America, Huntley and Birks (1983) for maximal rates in Europe, and Birks (1989) for average rates in Britain. The data are shown in Figure 4.11, using a logarithmic scale for rates. The American averages are rather lower than the British because they are taken over a longer period. Taking the two together suggests that 200 m per year is a typical average speed. Maximal speeds can be an order of magnitude greater than that, 2 km per year, but a typical maximum is about 600 m per year. (That column in the histogram represents 476 to 673 m yr^{-1}.)

How were these speeds achieved? The most direct answer is by Webb (1966), 'I frankly cannot imagine how this was done'. The difficulty is not just the distance per year but the number of years from one generation to another. But as all the flora, not just the trees, moved north, any native plant of either Canada or Britain seems to be capable of a sustained speed of around 200 m per year. Huntley (1993) discussed seven hypotheses for the extremely rapid

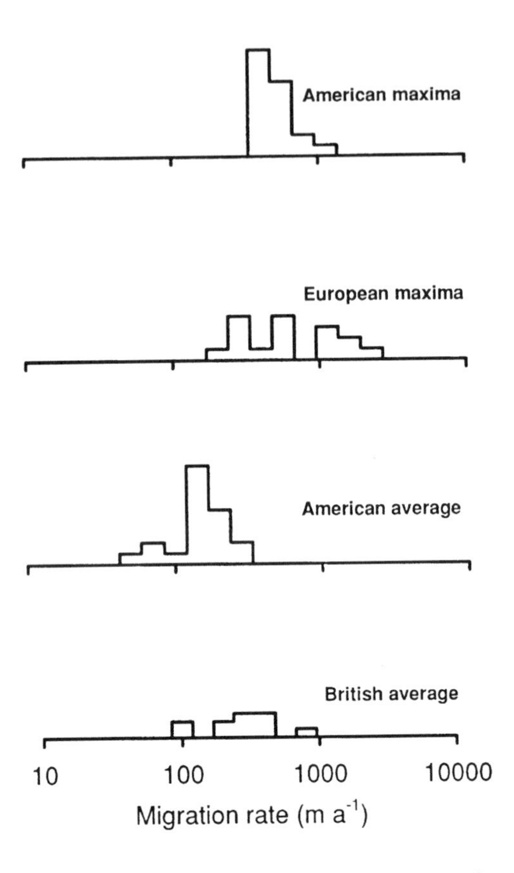

Figure 4.11 Histograms of the rates of spread of trees, on a logarithmic scale, in metres/year. American maxima from Delcourt and Delcourt (1987); European maxima from Huntley and Birks (1983); American averages from Delcourt and Delcourt (1987); British averages from Birks (1989).

spread of *Corylus* hazel. It is useful that he gives good reasons for thinking that neither refugia nor human transport can have been important in the mesolithic (ca. 10 000–5000 years bp). Wiggles in the radio-carbon calibration rate may be involved in some of the extremely high rates, but that cannot be much a factor for most of the rates shown in Figure 4.11. His other hypotheses are to do with the difference between *Corylus* and other genera, reactions to soil, climate and the succession of plant communities. How a large edible nut such as the fruit of *Corylus* moves several kilometres every generation is not addressed.

Some of these genera, *Betula* birches and the conifers for instance, have wind-borne seeds, and it is possible to imagine strong gales to be frequent enough everywhere to move a few seeds once a generation, or even more rarely if they move the seeds far enough. The real problem is with the heavy nuts of, for example, oaks and hazels. The most plausible and frequent suggestion (Johnson and Webb, 1989) is that birds such as jays (*Garrulus* in Europe, *Cyanocitta* in America) carry the nuts. But nuts of several species must be carried several kilometres each decade or so, and were it not for the evidence from the trees, the suggestion that jays ever behave in this way would be dismissed. Yet the seeds must be carried somehow. Maybe it just shows how little is known of the frequency of extreme behaviour whether by jays, squirrels, deer or other seed lovers.

With the analysed cases of section 4.3.4, the evidence that species typically spread faster than Fisher–Skellam theory indicates was weak, and in 4.3.5 it seemed that the analysed cases might be biased. The evidence from trees swings the argument back again. For plants, it seems that all native temperate species spread faster than ordinary small-scale observation and experiment would suggest.

4.5 SPREAD WITH ECOLOGICAL INTERACTION

The final example of spread leads into the subject of the next chapter, interactions between native and invasive species. It is the spread of the grey squirrel, *Sciurus carolinensis* in Britain and its interaction with the native red squirrel (Figure 4.12). *Sciurus vulgaris* the (Eurasian) red squirrel is a charming animal found across the palearctic from Portugal to Kamchatka (Figure 4.13). It is a tree squirrel, and primarily an animal of the boreal coniferous forests, but is also found commonly in deciduous forests and places with fewer trees, such as suburban gardens. At high population densities it can be a pest, particularly by attacking young conifers but also hardwoods, stripping them vertically

Figure 4.12 Red and grey squirrels, *Sciurus vulgaris* and *S. carolinensis*. The red, smaller with pointed and tufted ears, is on the left. The head and body lengths are about 22 and 26 cm, the tail lengths about 18 and 22 cm. (© Blackwell Scientific Publications, for the Mammal Society.)

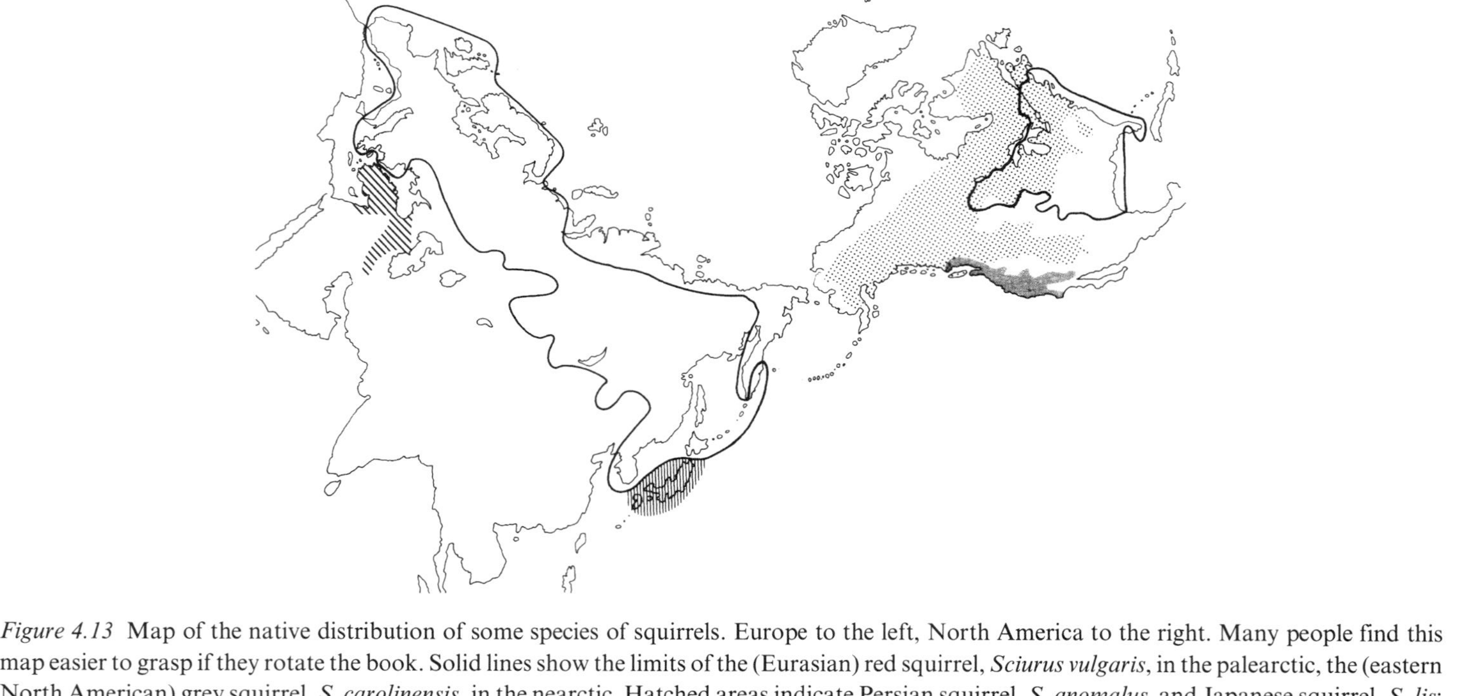

Figure 4.13 Map of the native distribution of some species of squirrels. Europe to the left, North America to the right. Many people find this map easier to grasp if they rotate the book. Solid lines show the limits of the (Eurasian) red squirrel, *Sciurus vulgaris*, in the palearctic, the (eastern North American) grey squirrel, *S. carolinensis*, in the nearctic. Hatched areas indicate Persian squirrel, *S. anomalus*, and Japanese squirrel, *S. lis*; stippled areas, American red squirrels, *Tamiasciurus*; coarse stippling, *T. hudsonicus*; fine stippling, Douglas' squirrel or chickaree *T. douglasii*. *Tamiasciurus* and the palearctic *Sciurus* are each sets of allospecies; there are no range overlaps. In the nearctic (depending on where the nearctic/neotropical boundary is drawn) there are at least another eight species of *Sciurus*. Most live in the west and in Mexico, but fox squirrel, *S. niger*, is largely sympatric with *S. carolinensis*. There are further *Sciurus* species in the neotropical realm (see Corbet and Hill, 1991). (Illustration by Mike Hill.)

which leaves scars in the wood, ring barking and damaging crowns. It will also eat fruit, flower bulbs, birds' eggs and nestlings.

The native range of *Sciurus carolinensis* (Figure 4.13) the (eastern North American) grey squirrel is the eastern United States west to about 97°W and from the Gulf of Mexico to Canada and into the southern parts of Ontario and Manitoba. It is a deciduous forest tree squirrel that adapts well to towns. Its range overlaps widely with the heavier and more terrestrial fox squirrel, *Sciurus niger*, and in the north with the American red squirrel, *Tamiasciurus hudsonicus*, which is primarily a coniferous forest species. All three species can be found in a zone from Maryland to Minnesota.

The grey squirrel can also be a charming animal. In America it is much hunted and seldom a pest. However, it Britain it is often a pest for the same reasons as the red squirrel, but more so. Its bark stripping of 10- to 40-year-old beeches, *Fagus sylvatica*, and other trees can be particularly annoying. As its population density is commonly much higher than red squirrels, it is more likely to cause damage. Because of this, and as an invader apparently driving out a native, it can be very unpopular. Its enemies often call it, inaccurately, a tree rat.

The grey squirrel has been introduced into many places, including cities in North America such as Vancouver and Seattle (Burt and Grossenheider, 1976), Cape Town in South Africa (Williamson, 1989a), Ireland at Castle Forbes in 1913 (Middleton, 1931), and to Piedmont and Liguria in Italy (Macdonald and Barrett, 1993). In Canada, pet squirrels frequently escape, and those and deliberate releases have led to populations in Saskatchewan, Quebec, New Brunswick and Nova Scotia, all north of the natural range (Peterson, 1966; Banfield, 1974). The origin and pattern of the ecological important spread in Britain is discussed below.

The rate of spread is variable. The North American urban squirrels are often confined to town parks and spread slowly if at all. The rate in South Africa was 0.4 km per year (Williamson, 1989a). In Ireland, where trees are often scarce, the spread was at a maximum of about 0.5 km per year, with no spread at all in some directions, judging from the map in Crichton (1974). In Italy, where no doubt there are many shot, the spread is only at about 1 km per year (from the data in Macdonald and Barrett, 1993), but those introductions are worrying for the future of red squirrels in deciduous forest, agricultural and suburban areas on the continent of Europe. In Britain, in some places and at some times, the population seems not to advance for some decades; in other places it goes slowly. The reason is apparently a lack of sufficient suitable habitat because of industrialization or agriculture. In East Anglia, Williamson and Brown (1986) estimated a mean rate of 7.7 km per year from the maps in Reynolds (1985) which covered 18 years of spread on a 5-km grid or module.

Grey squirrels have been introduced into Britain about 30 times, including three times in Scotland. The earliest successful one was in Cheshire in 1876. However, the bulk of the spread came from introductions in the south of England derived from a stock at Woburn, north of London, established in

Table 4.2 Quotations about the relationship of red and grey squirrels in Britain

Ritchie, 1920:
'The pets of confinement turn out too often to be the pests of the open, and already many plaints have been raised regarding the mischief wrought by these [grey squirrel] aliens, which Sir Frederick Treves protested in 1917, drive out our Red Squirrel'

Boyd Watt, 1923:
'the grey squirrel can never become wide-spread and dominant like our other introduced animals, the rabbit and the brown rat. . . . If a comparison has to be made . . . make it with the fallow-deer, a park and woodland animal, an attractive and valuable addition to our fauna'

Theobald, 1926:
'Wherever it [the grey squirrel] goes it drives out the Brown [= Red] Squirrel and in several districts in Kent, Surrey and Sussex where it is found the Brown Squirrel has become extinct, in one case it was erased in five years'

Middleton, 1931:
'The fact that the red squirrel has gone down in numbers during the past thirty years is therefore quite a different subject from its relations with the grey squirrel, since the latter can have little to do with the reduction of the native squirrels'

Sandars, 1937:
'There has been a marked decrease [of the red squirrel] since 1900, due to the introduction of the Grey Squirrel and other unknown causes'

Elton, 1958:
'The grey squirrel has replaced our native red one . . . it has taken up to fifteen years . . . We have no notion what happens'

Tittensor, 1977:
'Declines are largely independent of influence from grey squirrel and main cause probably widespread habitat destruction'

Gurnell, 1991:
'Reasons for population declines unclear; habitat destruction and disease may have been partially responsible but direct influence of grey squirrel not likely to be a factor. (Research required on possible indirect influences of grey squirrels on red squirrels . . .)'

Kenward and Holm, 1993:
'Since its first . . . introduction from North America in 1890, the grey squirrel (*Sciurus carolinensis*) has displaced the native red squirrel (*S. vulgaris*) . . . replacement of red squirrels can be explained by feeding competition alone'

1890. From that stock, squirrels were taken to many places, making it difficult to be sure how many introductions there were. Within a few decades they were unpopular pests and blamed by some for the decline and disappearance of reds. Whether greys are driving out reds has long been disputed, but it is now generally accepted that they do. Some quotes showing the disagreement are given in Table 4.2.

One reason for the prolonged disagreement is that red squirrel populations in Britain often die out. Indeed, both the present Scottish and Irish red squirrels come from introductions from England and, possibly, Europe. The Scottish populations, except possibly for a remnant in the Cairngorm mountains, died out by the beginning of the 18th century. They were reintroduced near Edinburgh in 1772, and in other parts of Scotland over the next century. In Ireland, where it is possible that they were never native, they were introduced at ten places, including Castle Forbes, in the 19th century. The details are in Shorten (1954) and the references she gives. In East Anglia, Reynolds (1985) noted red squirrel populations dying out anything from 18 years before to 16 years after the arrival of greys.

With so many points of introduction in England, the pattern of spread of the grey was complex. Lloyd (1983) attempted to map them. By the first survey of reds in 1945, reds were extinct in most of south-east England, apart from East Anglia, and also in parts of Cheshire, Yorkshire and south-west England. Greys by that time were found not just in those areas but also through the midlands and into Wales; and again in parts of southern Scotland. Now, greys are throughout almost all of England and Wales apart from a few islands and the far northern counties. There is considerable concern about the spread of greys in Cumbria (Lowe, 1993; Skelcher, 1993) and Northumbria, and the possibility that the Scottish and English populations will meet and fuse. Recent indices for England, Scotland and Wales were given in Usher, Crawford and Banwell (1992).

The reasons that an interaction was not accepted by many professional biologists for so long seem to be two. The first is that the mechanism was not apparent. Neither disease nor direct aggression seemed to be involved (see the Elton and Gurnell quotes in Table 4.2). The second is that as British red populations do die out spontaneously, and with the tendency of some scientists to adopt single explanation hypotheses, interaction was discounted. However, the pattern of spread and retreat encapsulated by the quote from Elton in Table 4.2, the persistence of reds on the Isle of Wight while the mainland opposite only had greys, the model of Okubo *et al.* (1989) and the observations of Kenward and Holm (1993) all contributed to the acceptance of interaction, and the initiation of programmes to save the red from extinction on the island of Britain.

Succinctly, the red is a less efficient forager in the deciduous woodland and anthropogenic habitats of Britain. It can not, for instance, digest acorns, while greys can. The result is both a lower r and a lower K. Okubo *et al.* (1989) modelled this interaction with a pair of reaction-diffusion equations

$$\begin{aligned} \partial n_1/\partial t &= D_1\nabla^2 n_1 + r_1 n_1(1 - b_1 n_1 - c_1 n_2) \\ \partial n_2/\partial t &= D_2\nabla^2 n_2 + r_2 n_2(1 - b_2 n_2 - c_2 n_1) \end{aligned} \qquad [4.6]$$

where, for $i = 1,2$,

$$\nabla^2 n_i = \partial n_i/\partial x + \partial n_i/\partial y$$

Table 4.3 Typical values for squirrels in low-density populations

Measure	*Red squirrel*	*Grey squirrel*
Typical weight (g)	267	500
Density (numbers ha^{-1})	0.75	10.00
Density (weight $g.ha^{-1}$)	200	5000
Litters ($year^{-1}$)	2	2
Survival		
First winter	0.5	0.5
Second winter	0.8	0.8
Young ($year^{-1}$)		
First breeding	2	3
Later	6	9
r*	0.61	0.82

*Intrinsic rate of natural increase.

and n_i is the population size, t time, D_i the diffusion coefficients, r_i the intrinsic rates of natural increase, b_i the self-limiting and c_i the competition coefficients.

The interaction here is competition, though the effect is almost entirely one way, the effect of the red on the grey is small. If that effect were zero, then the interaction would be defined as amensalism. Okubo *et al.* estimated parameters from the literature, and these, apart from the diffusion coefficient, are summarized in Table 4.3. The equations are difficult mathematically, but it is possible to show that there is likely to be a travelling wave solution. Simulation showed one, and some snap-shots of the replacement process are shown in Figure 4.14. Note that the parameters lead to a correct prediction that reds will become extinct between 10 and 20 years after the arrival of greys. With this model, the advance of the grey is essentially the same as the advance from equation [4.2] except that it will be slower because of competition. The retreat of the red will be at the same speed as the advance of the grey.

The estimate from the literature of the diffusion coefficient of the grey squirrel, D_g, was only 1.25 km^2 yr^{-1}. That would only give an advance of about 2 km yr^{-1} without competition, compared with an observed 7.7 km yr^{-1} with competition in East Anglia. Once again, the diffusion coefficient estimated from local short-term movements seems too small. Okubo *et al.* sketched out a model with squirrels moving between isolated woodlots which would fit the observations.

The general conclusion is that reaction-diffusion equations, basic linear deterministic equations without complications such as age structure, give a satisfactory model of spread with competitive interaction. This study encapsulates the main conclusions of this chapter. The Fisher–Skellam equations, equations [4.1] and [4.2], give a good first-order description of spread in many cases. The major weaknesses are first, that the calculated rate of spread is often less than the observed and second, that it has not yet been possible to model the geographical variation in rates of spread seen in so many cases. But

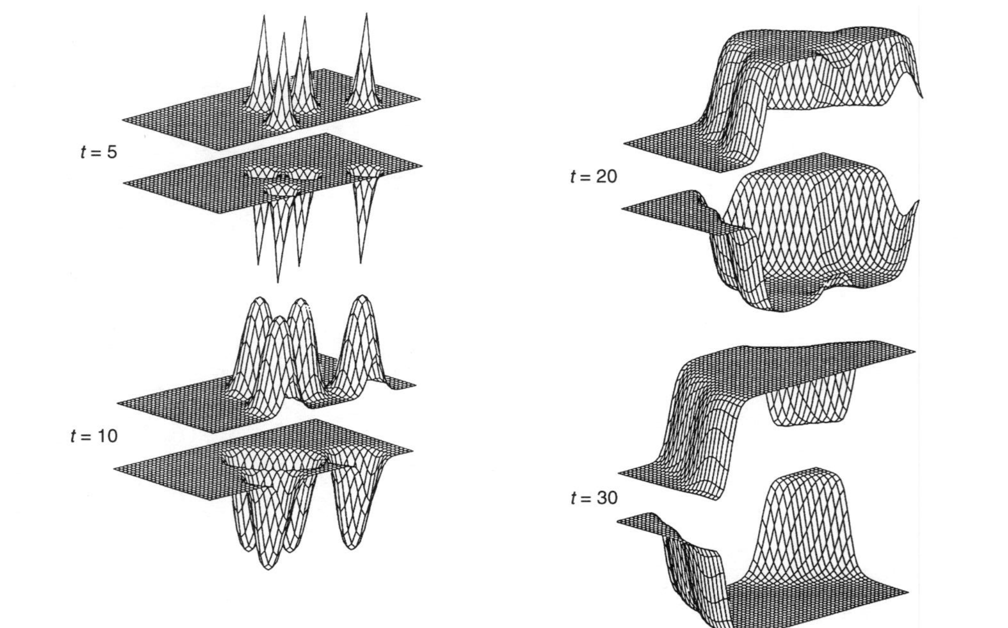

Figure 4.14 Three-dimensional surfaces of the numerical solution of equations [4.6]. At the start, greys are uniformly absent, a plane of zero density, while reds are uniformly present, a plane of 100% density shown, for clarity, below the grey plane. Greys are introduced at four points at time $t = 0$. The figures show the increase and spread of greys, and the retreat of reds, from these four points at times 5, 10, 20 and 30. Greys almost reach 100% at $t = 10$, coalescing and spreading at 100% thereafter. Note that reds start decreasing at the same time and same speed as greys are increasing, are about to go extinct at four points at $t = 10$, are extinct over a continuous area at $t = 20$, and suffer a travelling wave of extinction thereafter. (From Okubo *et al.*, 1989.)

there is still much to be learnt about spread, both empirically and theoretically. We need geographically variable models, with functions describing how r and D vary spatially. We need empirical observations that will explain the post-glacial spreads. We need to know more about how frequently the model predictions are too slow, too fast, or of the wrong form.

Spread is the middle stage of invasions. What happens after the spreading phase has been completed is the main theme of the rest of the book.

5 Ecological consequences of invasions

Conceptual framework points (from Table 1.1):

CFP 6 Most invaders have minor consequences (tens rule)
CFP 7 Major consequences have as
•effects: depressed populations to individual extinctions to ecosystem restructuring
•mechanisms: enemies (vertical food-chain processes); competition, amensalism, swamping (horizontal food-chain processes)
CFP 10 Invasion studies are relevant to considering the risks of introducing new species or genotypes, the release of genetically engineered organisms and the success and consequences of biological control

5.1 INTRODUCTION AND MINOR EFFECTS

The tens rule (section 2.3) says that 10% of introductions, of casual species, will become established or naturalized, and that 10% of those will become pests. The rule is a very rough one, and pests are defined by human perception rather than by ecological effects. But the rule does have the implication that most established introductions will have only minor ecological effects, and this seems to be true. Unfortunately, the major effects can be very large indeed. In trying to predict possible damage from invaders, we are trying to foresee events of low probability but high impact, and that is inevitably very difficult.

One way of making this easier is to have some idea of the types of damage that have been important in the past. In this chapter I will consider ecological effects, in the next, Chapter 6, genetic effects. I will consider surveys of effects, and some examples of minor effects, and some difficulties in being sure they are minor. With major effects there are so many examples, many famous, that I can review only a few. The most important major effect is probably being an enemy of the native biota, by being a predator, parasitoid, parasite, pathogen or herbivore. Sitting on top of the food-chain, or being a generalized enemy, an enemy of many species, seem particularly likely to produce marked ecological effects (Pimm, 1991). Another important effect is swamping, growing in dense concentrations or clumps that exclude other species. Competition is apparently much less often an important effect, and certainly much harder to prove. Moving on from these single effects, some invaders have what can be called ecosystem effects, though the distinction is fuzzy. The chapter ends with a discussion of some extinctions resulting from invasions. Extinction of a whole species is for ever, an irreversible effect and a major subject in its own right. It is treated further in Chapter 7.

Invaders can be said to have a small probability of producing a large effect. The small probability was discussed in Chapter 2; some of the large effects in a recent survey by the United States Congress Office of Technology Assessment were also mentioned there. Using criteria that could no doubt be challenged but are clearly explained, that report (US Congress OTA, 1993) estimated the cumulative cost to the US economy of what it called Harmful Non-Indigenous Species to have been almost a hundred billion dollars, US$ 96 944 000 000. That is not only a very large effect, it is one that has been steadily increasing, and is still increasing, each year's cost being greater than the one before. The yearly cost just in the USA is now probably well over a billion dollars for the human costs and benefits. The pure environmental aspects are scarcely considered.

Most of this cost comes from insects, including gypsy moth, *Lymantria dispar*, the fire ants, *Solenopsis invicta* and *S. richteri*, and many others. The insects are 43 of the 79 species analysed (54%), but their cost is 96% of the total. At least another 478 species, 69% of them insects, are known to have harmful effects. From the general rule that most species have minor effects, costing these would probably not increase the total cost greatly. The report says, disarmingly, 'information was incomplete, inconsistent, or had other short comings for most of the 79 species'. But however the figures might be adjusted, there can be no doubt that introduced species have had, world-wide, colossal costs both economic and ecologic. Even the smallest cost in the report, US$ 225 million for six terrestrial vertebrates, is a large sum. Thirty-nine other such species were not analysed.

How many insects have been introduced altogether into the continental US? Sailer (1983) counted 1554 species, and this count is now updated and kept on a USDA database (Simberloff, 1989). It is very likely that the count is appreciably low, because of species not found, incorrectly identified or not recognized as introductions. A figure now of between 2000 and 2500 would seem likely. In Hawaii (not included in the figure) it is known that at least 20 new species establish each year (Anon., 1994a), and about 2000 invertebrates, mostly insects, may well be established already (Howarth and Medeiros, 1989). The massive disruption of habitats, and possibly being an oceanic island (see section 2.3.2), make Hawaii more invasible, but it is also a very small area. The Hawaiian figures reinforce my view of the continental figures. But even if that is not accepted, the majority of the established continental US-introduced insects are not known to cause harm, and would

Plate 1 *Veronica filiformis*. The flowers are about 1 cm across. (Photograph by Michael Usher.)

Plate 2 Two Atlantic fulmars *Fulmarus galacialis*. The birds have a wing span of about 1 m. The generally gull-like appearance and the special beak structure with its tube-nose can be seen. (Photograph by Gordon Woodroffe.)

Plate 3 A rabbit *Oryctolagus cuniculus* suffering from myxomatosis. The swollen head and closed eye is typical of the disease. The body length is usually 40–50 cm. (Photograph by Michael Usher.)

Plate 4 Flowers of *Impatiens noli-tangere* (yellow) and *I. glandulifera* (pinkish purple). The former are about 2 cm long, the latter about 3 cm. (Photographs by Michael Usher.)

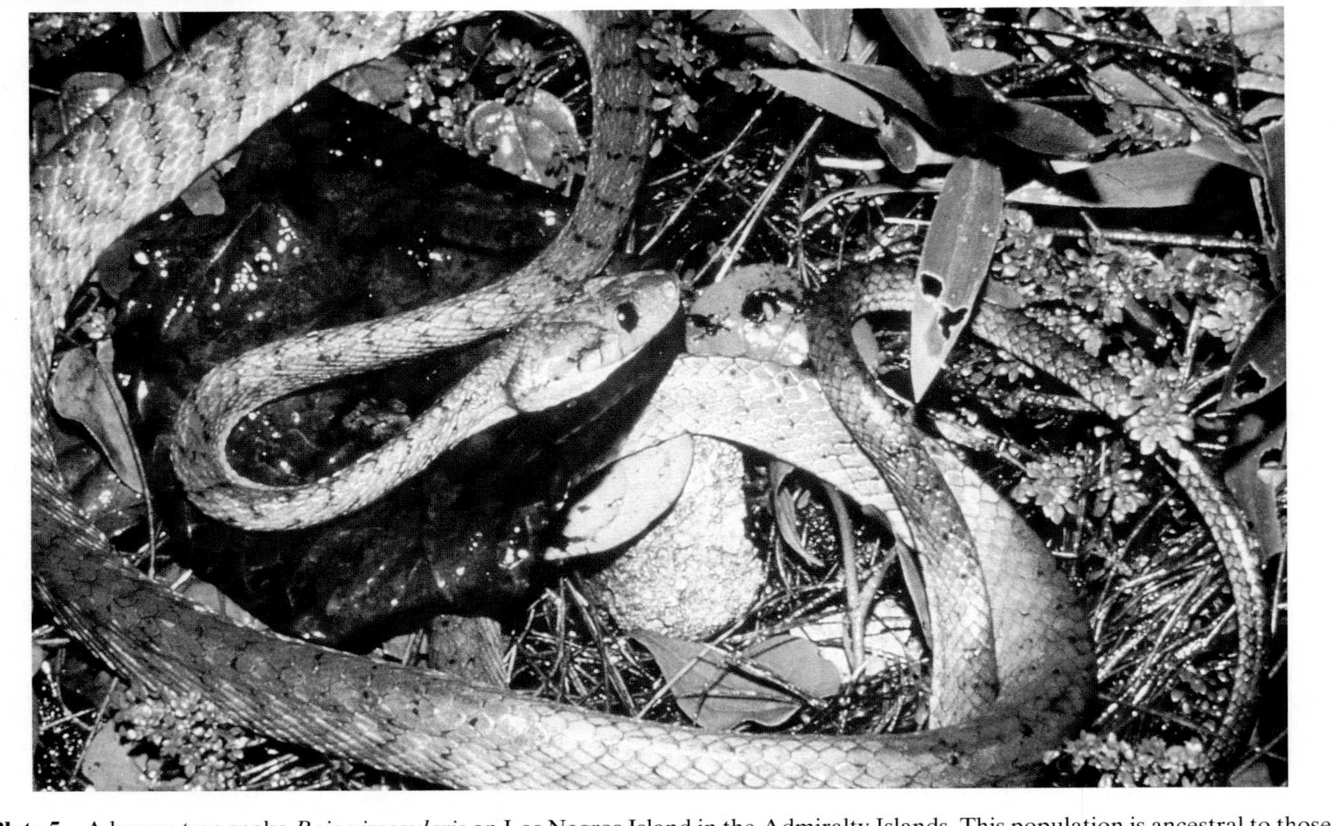

Plate 5 A brown tree snake *Boiga irregularis* on Los Negros Island in the Admiralty Islands. This population is ancestral to those imported to Guam. A fully grown snake is about 2 m long. (Photograph by Gordon H. Rodda, National Biological Service (U.S.).)

Plate 6 *Euglandina rosea*, a predatory snail, and its reflection in glass. The shell is about 3 cm long. (Copyright © BBC.)

Plate 7 Africanized and European worker bees *Apis mellifera* in the same colony. The smaller, darker bees are Africanized. This mixed colony was created by requeening an Africanized colony with a European queen. The spacing of the comb cells is 4–5 mm. (Photograph by Mark Winston.)

Plate 8 A sward of *Spartina anglica*. Plants can grow to 130 cm, but these are rather less than a metre high. (Photograph by Alan Gray.)

appear to have minimal ecological effect, though most have not been studied in detail.

Welcomme (1992) surveyed established introduced freshwater ('inland aquatic') species, 291 species in 148 countries. His correspondents reported 49 species with negative effects, 45 with positive ones and 14 equal (equally positive and negative). The remaining 183, following CFP 6, had 'insufficient environmental effect to warrant mention'. Note that while the negative effects are at the high end of the tens rule, the definition of negative was '[had] not met the objectives for which they were introduced or had shown sufficient environmental impact to be pests'. Taking individual introductions, rather than summing over species, gives 9.1% negative, 1.0% equal, 12.5% positive and 77.4% no comment, 1673 introductions in all. The equal are 18 species 'whose introduction is viewed with mixed feelings'. Three of them have the characteristic of many pest fish species of forming dense, stunted, populations, but six are predators. One of those predators is *Cichla ocellaris* discussed in section 5.3.1.

General accounts of invasion often give the impression that marine invaders are scarcer than terrestrial ones. In fact the data are not available for a quantitative comparison. As on land, most marine invaders have only minor effects. *Biddulphia sinensis*, a diatom (see Figure 3.4), was discussed in section 3.4.6. A few other examples of marine invasions are listed in section 4.3.4. Like *B. sinensis*, some have minor but detectable effects in their communities. The barnacle, *E. modestus*, for instance, sometimes takes space from the common European *Balanus balanoides*. Indeed, by suppressing small *Balanus*, the remaining ones sometimes individually are rather large and healthy (Williamson and Brown, 1986). The spread of *E. modestus* was characterized by a mixture of gradual advance and occasional leaps, a phenomenon that may be common in the sea (section 4.3.4). It is notably tolerant of low-salinity sites around the high water mark, and so occupies zones further up estuaries than other barnacles in Britain (Crisp, 1958).

That marine invaders are very common is shown by the study by Carlton and Geller (1993) of the fauna and flora of ballast tanks mentioned in section 2.2. Table 2.1 lists a selection of some 45 successful invaders world-wide, including *Dreissena polymorpha* and *Mnemiopsis leidyi*, discussed below. The species found in ballast tanks over 5 years at Port of Coos bay, Oregon, USA, included 72 species of crustacea, 43 of annelids, 11 of protozoa and 128 species of diatoms. Those are species that arrived; it is not known how many had already or were likely to become established.

Before going on to serious consequences of invasions, there are two more points to be made about invasions with minor consequences. The first is that a species can be common and yet have little measurable effect on other species. The second is that it may take a long time for a species to change from innocuous to pest, and so to that extent, estimates of how many invasions are harmless may, in the long run, turn out to be underestimates.

The fulmar (section 1.3.1) is an example of a bird now common which has scarcely affected the communities it has invaded, whether of cliff-nesting birds or, apparently, the marine life it feeds on. A newer example is the carabid beetle, *Pterostichus melanarius*, which reached Alberta, Canada, from Europe about 1959, and is spreading out from Edmonton. It is now one of the ten most common predatory ground beetles in central Alberta, but with no measurable effect as yet on other species (Niemalä and Spence, 1991, 1994).

A third example of the small effect of common species is the set of breeding wildfowl (ducks, swans and geese, Anseriformes) in the British Isles (Table 5.1). A surprisingly high proportion have invaded naturally. That several species have been introduced and escaped is less surprising, as these birds are popular with aviculturists. Only two have caused concern. Canada geese,

Table 5.1 Breeding wildfowl of the British Isles. Reliable records of status go back to the 18th century, so it is possible to divide the present-day breeders into those that were breeding in the British Isles then, and those that invaded naturally since then or were introduced. Species with dual status appear in two columns, as do those that have become extinct (e) and have re-invaded. The order is that used in Gibbons, Reid and Chapman (1993).

Species	*Native*	*Date of natural invasion*	*Date of non-captive breeding*
Mute swan, *Cygnus olor*	x		
Whooper swan, *C. cygnus* (established?)	xe	1978	
Greylag goose, *Anser anser*	x		1959 & pre
Canada goose, *Branta canadensis*			Pre-1840
Egyptian goose, *Alopochen aegiptiacus*			18th century
Shelduck, *Tadorna tadorna*	x		
Wood duck, *Aix spousa*			1969
Mandarin, *A. galericulata*			1928
Wigeon, *Anas penelope*		1834	
Gadwall, *A. strepera*		1850	
Teal, *A. crecca*	x		
Mallard, *A. platyrhynchos*	x		Often
Pintail, *A. acuta*		1869	
Garganey, *A. querquedela*	x		
Shoveler, *A. clypeata*	x		
Pochard, *Aythya ferina*	x		
Tufted duck, *A. fuligula*		1849	
Scaup, *A. marila* (established?)		1897	
Eider, *Somateria mollissima*	x		
Common scoter, *Melanitta nigra*		1855	
Goldeneye, *Bucephala clangula*		1970	
Red-breasted merganser, *Mergus serrator*	x		
Goosander, *M. merganser*		1871	
Ruddy duck, *Oxyura jamaicensis*			1960

Branta canadensis, have become very numerous in the last few decades, and are now sometimes an agricultural pest and more frequently a social nuisance from fouling paths. Ruddy duck, *Oxyura jamaicensis*, are not a problem in Britain, but a conservation worry elsewhere for genetic reasons, discussed in section 6.3.3. The other dozen or so have become established without important ecological effects.

Canada geese had been around for more than a century, and ruddy duck for some decades, before either caused concern. These sort of delays are common, and may make statistics on the low proportion of pests too optimistic. Another example (section 1.3.3) is *Impatiens glandulifera*, which was declared a weed in 1890, 35 years after it was first recorded in the wild, and 51 years after it was first imported (Perrins, Fitter and Williamson, 1993). Even now it is often not considered a pest, as it can be a colourful addition to the flora along canal banks. Like many other species, it is a pest in some places, particularly woodlands, but not in others.

Another species which was thought to be benign for a long time is the muntjac deer, *Muntiacus reevsi*, in England. It is a Chinese species, and was first released from Woburn in Bedfordshire in 1901. It is the smallest deer now in Europe, with short simple antlers, and is a browser of woodland for the most part. Although reproduction is more or less continuous, females average only slightly more than one fawn a year (Chapman, 1991). Its natural rate of spread is about 1 km per year, but it has frequently been introduced into other parts of Britain, so that its apparent rate of spread is much greater (Chapman, Harris and Stanford, 1994). It is now found in most parts of England, some places in Wales, a few in Scotland, and on the Isle of Man, but only at high density in south-east England, near the original sites of introduction.

Only in the 1990s has it become regarded as a pest. Digging for underground plants damages vegetation, and browsing on coppice interferes with woodland management. It is especially a pest, 'devastating', for gardeners. Muntjac can cause serious losses in market gardens and orchards (Chapman *et al.*, 1994). Steps towards this state are shown by some quotes. 'Appears to be a harmless introduction' (Dansie, 1977); 'The animals are an almost innocuous asset to the countryside and they give pleasure to thousands and pain to few' (Dansie, 1983); '. . . very little if any damage is done to conifers or arable crops. . . . The shoots of some deciduous trees and some garden plants, e.g. roses are more vulnerable. . . . Errant muntjac in a town can create hazards.' (Chapman, 1991). No doubt its slow reproduction and slow spread led to the slow recognition of the problems with this species.

One introduction that Elton (1958) approved of was of the multiflora rose, *Rosa multiflora*, into North America. It appeared to him to have only benefits in improving agricultural landscapes. However it spread vigorously on marginal land, including rights-of-way. It has since 1967 been declared a noxious weed in several states. Cutting, burning, bulldozing, herbicides and goats all fail to control it. The search is on for a specific biocontrol agent, but with so many species and varieties of *Rosa* in cultivation and the wild, the search is difficult (Amrine and Stasny, 1993).

Even species recognized as pests may become much more severe pests with time. *Mimosa pigra* is a spiny South American shrub introduced to Australia sometime between 1870 and 1890. It can grow 5 m tall, with a density of one plant per m^2, and was a minor but troublesome weed for about a century. In the late 1970s there was a major increase in the Northern Territory (Braithwaite *et al.*, 1989). It now occurs in dense stands affecting 450 km^2, displacing some of the original communities. Alongside the billabongs there are sedgelands, dominated by *Eleocharis spiralis*. Bordering them are the paperbark forests with a mixture of trees including *Melaleuca cajupti* and *M. viridiflora*. Billabongs are the old beds of rivers, oxbows, and had little or any macroscopic vegetation. *Mimosa pigra* takes over all of the sedgelands, most of the billabong and a great deal of the paperbark forest, creating a new, impoverished, ecosystem. This is shown diagramatically in Figure 5.1.

With this example, we have moved from benign introductions to damaging ones, from minor effects to major.

5.2 BENEFITS, COSTS AND PERCEPTIONS

Before going on to look at some of these major ecological effects, enmity in section 5.3, competition in section 5.4, ecosystem effects in section 5.5 and extinction in section 5.6, there is a different dimension to consider. That is, how to judge these ecological effects. Are any of them beneficial? Can there be agreement on the damage, and so the remedial actions needed, of the harmful ones? Biological control is often claimed to be entirely beneficial, and some aspects were discussed in sections 2.3.2, 3.3.1 and 3.4.3. Section 5.2.1 gives an overview of the technique, and an assessment of the circumstances in which it can be ecologically harmful. However, there can be conflicts between ecology and economics, or between different economic interests, and these are considered in section 5.2.2.

As most of the examples in this book are of harmful invasions, it is as well to remember that invasions can sometimes be beneficial. Few invasions cause extinctions, and so most invasions lead to an increase in local biodiversity, just counting the species present. In derelict and impoverished areas this can be a benefit. For instance, *Impatiens glandulifera* (sections 1.3.3 and 5.1) can improve the appearance of urban areas, even though it is a weed elsewhere. The carabid beetles that have been introduced into arctic areas (Table 2.7) may have brought ecological benefits, and the communities are still impoverished. In geological time, invasions have been an important driving force in evolution, a point I return to in section 7.4.2.

5.2.1 Biological control

There are many styles of biological control, of using biological agents to control undesirable species. Two extreme types can be called classical and

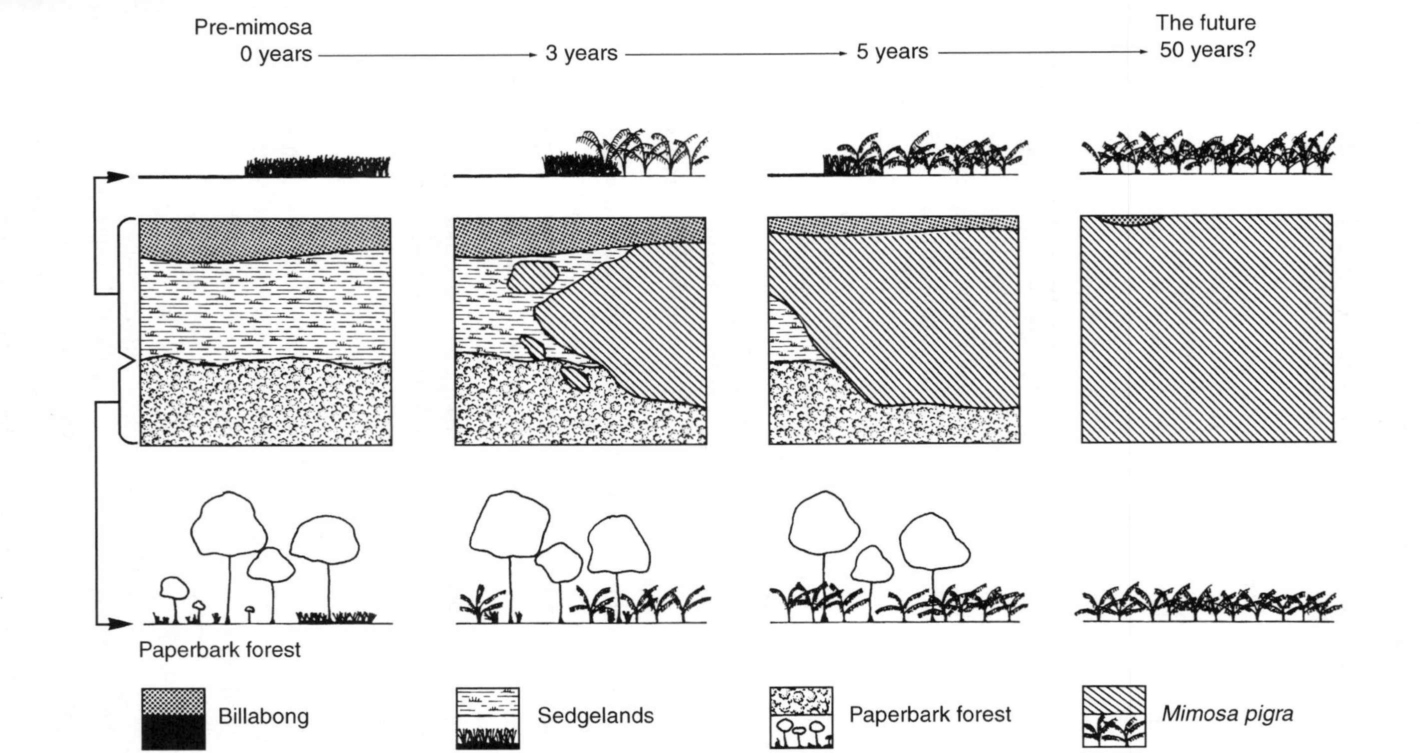

Figure 5.1 The spread of *Mimosa pigra* in valleys in Northern Australia. The central line of figures are diagrammatic maps, the drawings above and below diagrams of the appearance of the vegetation at each stage. (From Braithwaite, Lonsdale and Estbergs, 1989, with permission of the authors and Elsevier Science Publishers.)

inundative. Classically, the undesirable species is an introduction that is a pest partly because it has been introduced without its natural enemies, and so has a high population density. It is either a weed or an arthropod that attacks crops. The strategy is to search the area of origin for suitable enemies, grow these enemies in quarantine to rid them of their enemies and to test the host range, and then to release species that are approved. The programme may need $2 million and 20 scientist-years (Harris, 1985). The intention is to establish one or more new species, which will then control the pest at no further cost. Ideally, both pest and control agent will persist at low densities.

Inundative control is using the biological agent as if it were a chemical pesticide, spraying it on the pest and getting an immediate kill. The advantage over ordinary chemicals is that the agent is specific, or at least more specific, and short-lived. *Bacillus thuringiensis*, a bacterium that produces a protein toxic to insects, is often used in this way. Some of the claims for specificity are surprising. The most extreme I have seen is that the agent only attacks invertebrates. However, claims to attack only lepidoptera, or coleoptera, are also put forward as claims of adequate specificity, which they are clearly not. Pure inundative control does not involve the establishment of species, but strategies intermediate between classical and inundative may.

Biological control almost always involves a predator, parasitoid or pathogen. Competition has been tried in the control of schistosomiasis. This human disease is caused by the trematode blood fluke *Schistosoma*, which has a complex life history that involves a stage in certain freshwater snails. Control of the disease can be effected by control of the snails by bringing in competitors (Pointier, Guyard and Mosser, 1989).

A more typical example, almost the only example of successful biological control in Britain, involves the use of an imported predator to control an imported herbivore on imported species of tree. The trees are spruces *Picea*, particularly Norway spruce, *P. abies*, from north and east Europe, and Sitka spruce, *P. sitchensis*, from western North America; there are no native British spruces. The herbivore is a scolytid bark beetle, *Dentroctonus micans*, which attacks healthy trees and has long been known as a pest of spruce in Europe. It was found in Britain in 1982. It occasionally attacks pines *Pinus*. The predator is another beetle, *Rhizophagus grandis*, which apparently can only breed when fed on *D. micans*, and which is specifically attracted to chemicals given off by the prey. As *R. grandis* had already been used and studied as a control agent in Europe, import and release licenses were quickly obtained, and the first releases were in 1984 (Evans and Fielding, 1994).

R. grandis has been successful in keeping down *D. micans* infestations, but the bark beetle is still spreading naturally at between 3 and 8 km yr^{-1}. So, the control strategy also involves preventing long-distance dispersal by timber transport. The evidence on specificity is scattered through the literature. The larvae can feed on other prey, and specificity depends on the behaviour of the adults, helped by all other British scolytids being much smaller than *D. micans* (Cooter, 1991; Evans and Fielding, 1994). This example has many features typical of classical biological control: an apparently specific enemy, which can survive at low population densities, a reduction but not elimination of the pest, the need for other measures.

Biological control is often put forward as entirely safe, with no environmental damage, and so much to be preferred to chemical control. It can be brilliantly successful, as in the control of *Salvinia* (section 5.4), or pointless and disastrous, as with *Euglandina* (section 5.6), though most releases fail to establish control (Table 2.6). There is a dispute about how safe biological control is (Samways, 1994; Simberloff and Stilling, in press). The disagreements are primarily about how often arthropod control agents attack non-target species. Regrettably professionals involved with the biological control of insects have normally only considered the economically important species (Waage and Greathead, 1988); species of conservation interest have been ignored.

The record with control of weeds is much better as specificity testing has been more stringent. However, not all weed biocontrol experts have had high environmental standards. Markin and Yoshioka (1992) had another (cf. section 3.3.1) scoring scheme for the likely success of biological control. While they accepted that a weed that is also a major agricultural crop elsewhere should never be targeted, they would have allowed related species, either of minor agricultural importance or endangered, to be attacked. They gave no consideration whatever to related species of purely ecological and biodiversity interest, ignoring the potential of biological control to make such species endangered or extinct. With such attitudes, it is not surprising that biological control is no longer viewed universally as desirable.

There is no dispute that biological control by vertebrates, such as mongoose *Herpestes auropunctatus*, has almost always been disastrous ecologically; they are not specific predators and attack many non-target species. There is also no dispute that a species specific biological control agent that remained specific and attacked a species that is always a pest would be acceptable.

The three main issues then are: (1) how specific must an agent be, (2) will a specific agent evolve to be less specific and (3) should the target species be controlled? Point (2) is dealt with in section 6.4.2, and point (3) implies conflicts of interest, which are discussed in the next section. Point (1) also involves conflicts of interest. The moth *Cactoblastis cactorum* from Argentina was enormously successful in clearing Queensland, Australia of the prickly pear cactus *Opuntia stricta*. Having spread round the world, it attacks many other species of *Opuntia*, which is a large genus. In the US, *C. cactorum* causes concern to some by attacking rare endemic *Opuntia*s (US Congress OTA, 1993). In South Africa (and elsewhere) *Opuntia* are used for stock fodder, human fruits and vegetables, for rearing insects to produce a red dye, and other uses (Samways, 1994). Objections to *Cactoblastis* arise entirely from its limited specificity.

The exception to the rule that biological control agents should be completely species-specific can be when the other species attacked are also pests, or, more dubiously, when the release is in an area where no other susceptible hosts occur. An example of the first is a case of an extinction which seems to have been entirely welcome, the citrus psylla *Trioza erytraea* on the island of Réunion in the Indian Ocean (Samways, 1988). As its English name suggests, on that island this homopteran is an imported pest on an imported crop. The introduced hymenopterous parasite, *Tetrastichus dryi*, drove *Trioza erytraea* to extinction, because it could maintain a population on an alternative host *Trioza eastopi*.

That is a simple example of how many extinctions occur. An attacking species with a few victim species can maintain itself on some, while eliminating others. This mechanism explains why conservationists are dismayed by the attitude of some agricultural entomologists. An agent controlling an agricultural pest could, in appropriate circumstances, drive species outside agriculture to extinction. This seems to have happened in New Zealand (Roberts, 1986) and probably in Hawaii (Simberloff and Stilling, in press). Funasaki *et al.* (1988), who doubt most of the Hawaiian cases, also doubt that *Euglandina* is undesirable. *Euglandina* is the snail responsible for mass extinctions on Pacific islands (section 5.6), so Funasaki *et al.* seem not entirely objective. The only demonstrably safe biological control agents are those that are completely species-specific.

With chemical control there can be much ecological damage. Some such damage could be permanent. Damage by biological control is almost guaranteed to be permanent.

5.2.2 Conflicts of interest

It is not unusual for an introduction to be welcomed by some interests, particularly those sporting and commercial, while condemned by environmentalists. Such disputes were evident in Welcomme's survey of the effects of fish introductions, discussed in section 5.1. Pigs on Hawaii are a conservation disaster, yet there are hunters who want to keep them (and animal rights activists who object to all usable methods of control). In general, there is no way that all parties in such a dispute can be satisfied. Economists would suggest a cost-benefit analysis, which could certainly be made to produce a numerical answer. As the costs and benefits accrue to different people, that would not end the dispute. A clear example of the impossibility of balancing different interests comes from Lake Victoria in Africa.

Lake Victoria is the largest tropical lake, 68 000 km^2, and until recently its fish fauna was dominated by cichlids. There was a most interesting swarm of more than 300 haplochromine cichlids, 99% of them endemic, exploiting virtually all the food sources in the lake. By weight, 80% of the demersal fish were haplochromines (Witte *et al.*, 1992).

Sometime between 1954 and 1957, Nile perch *Lates niloticus* were placed surreptitiously in the lake, though they were not detected until 1960 (Reynolds, Gréboval and Mannini, 1995). Nile perch is a large predatory fish, growing to 180 cm, and is a primary but not the only cause of a drastic change in the ecology of the lake. As so often, its effects were not noticed for some time, but from about 1980 or a little before, landings increased rapidly. The total catch of all fish in the lake was about 100 000 tonnes in the 1970s. By the end of the 1980s it had increased to about 500 000 tonnes, 300 000 of them perch. The rest of the catch was predominantly the introduced Nile tilapia, *Oreochromis niloticus*, and the native *Rastrineolobea argentea*, neither of which had been important in the catch in the 1970s. The latter is a pelagic zooplanktivorous cyprinid, called dagaa or omena.

At the same time, about 200 of the haplochromine species disappeared, and it is thought that most of them are extinct. 'A unique group of vertebrates has been lost and most of the other indigenous fish species (including the remaining haplochromines) have shown a strong decline. The formerly diverse icthyofauna is now dominated by only three fish species.' (Witte, Goldschmidt and Wanink, 1995).

Part of the changes come from increased fishing pressure and the use of finer nets. One reason that dagaa was not important, but now is, is that it is only caught by mesh sizes below 10 mm, used now but not earlier. All management recommendations have been ignored (Bundy and Pitcher, 1995). There have been marked fluctuations in the water level and the lake has become more eutrophic. Algal biomass has increased two- or three-fold, and a deep anoxic layer has developed and is increasing (Oguto-Ohwayo, 1995). Nevertheless, there is now no doubt about the impact of predation by Nile perch (Bundy and Pitcher, 1995).

In biological and environmental terms, this invasion has been a disaster. In terms of feeding the growing human population round the lake it is a success. There is no way to reconcile the conflict of these views. Conflicts of interest also occur on land.

In biological control, there can be disputes not only between environmentalists and agriculturalists, but between different types of agriculturalist. In Australia, *Echium platagineum* is known as Paterson's curse to farmers, as it takes over grazing land. Biological control was proposed. When it flowers, large areas turn purple (Delfosse and Cullen, 1985), and it is valued by bee-keepers who call it Salvation Jane. The bee-keepers therefore opposed biological control. There was no simple legal way to settle this dispute, so a mechanism was provided in the Biological Control Act (Commonwealth of Australia, 1984). The Industries Assistance Commission, one of the bodies that can consider applications under the Act, approved eight insect species (Cullen and Delfosse, 1991). Some of these could possibly attack native Australian species. The emphasis in the submission was on the economic importance of plants. The Act requires that 'the release of the relevant organisms would not cause any significant harm to any person or to the environment, other than the harm (if any) resulting from the control throughout Australia of target organisms'. The interpretation of 'significant' would be a matter for lawyers. So far, none of the agents tried has been particularly effective (R. Groves, personal communication).

Other conflicts of interest in biological control between farmers on the one hand and bee-keepers, gardeners, environmentalists and other farmers on the other are described by Syrett, Hill and Jessep (1985) for New Zealand and Turner (1985) for USA. The problems are both specificity and whether the species should be controlled at all.

In all these cases steps can be taken to get as objective a view as possible, but in the end decisions about whether an introduction is desirable will be value judgements. Environmentalists will want to apply the precautionary principle, of which one statement is (Dovers and Handmer, 1995): 'Where there are threats of serious or irreversible environmental damage, lack of scientific certainty should not be used as a reason for postponing measures to prevent environmental degradation. In the application of the precautionary principle, public and private decisions should be guided by: (i) careful evaluation to avoid, wherever practicable, serious or irreversible damage to the environment; and (ii) an assessment of the risk-weighted consequence of various options.'

5.3 CONSEQUENCES OF ENMITY

Being an enemy, which here means eating or consuming other species, is the most likely way that an introduced species will have a major ecological effect. Simberloff (1981) reviewed the literature on introductions. As he was testing the MacArthur and Wilson theory of island biogeography, he was looking for extinctions, one-to-one replacements. The criticism that he ignored major effects not leading to extinction (Herbold and Moyle, 1986; Pimm, 1989, 1991) is simultaneously true and misdirected. Simberloff was using records of introduction to study a biogeographical theory; his study was not a wide survey of the effects of invasions. As he points out, the cases reported were undoubtedly biassed to the more striking invasions.

But Simberloff's counts of the numbers of different types of invasions causing extinction are a useful pointer to the proportion of different types of major effects caused by invaders. He noted 71 extinctions, 50 caused by predation, 11 from habitat change and only three from competition, which leaves seven not ascribed. While that may exaggerate the effect of predation, its importance, and the rather minor impact of competition, fit most of what we know about invasions. The examples in Chapter 1 fit that. Rabbits have major effects from grazing, but minor ones from burrowing and no effect which has been called competition, though obviously other organisms eating the same plants must have been affected. Myxomatosis is, in this context, a micro-predator, while the major effects of *Impatiens*, swamping other species, is amensalism rather than habitat change or competition. More major swamping effects could be called habitat change.

With so many examples of the high ecological impact of enemies, only a few can be mentioned in this section; there are further examples in section 5.6 on extinction. I will consider macro-enemies, that is, predators and grazers, and also parasitoids, before micro-enemies or pathogens.

5.3.1 Predators and grazers

The first examples are some well-known terrestrial species, illustrating the sort of ecological damage possible from enemies. Predators may induce predator–prey cycles, and take some time to come to equilibrium. There can be a boom-and-bust as discussed in section 2.4.1. Introduced predators can interact in complex ways with the native biota, with other introductions and with other environmental stresses such as pollution and eutrophication. Two aquatic examples in this section show the difficulties of disentangling cause and effect in such situations. Two more predators, one introduced accidentally, one for biological control, are the villains of section 5.6.

Perhaps the most notorious immigrants are certain mammals. Cats, rats, goats and rabbits have particularly damaged island biota. They come in at the top of the food-chain, and often meet organisms that have no defence against their attack. Ground-nesting birds have often been the victims of cats and rats. In New Zealand, introduced mammalian predators such as dogs and stoats, *Mustela erminea*, along with habitat destruction, are the prime cause of the decline and extinction of many native birds, but cats have 'by far the worst record' (King, 1984). The cats are feral domestic cats, *Felis catus*, a species derived from and interfertile with *F. silvaticus*, a widespread species with several subspecies in Europe, Africa and Asia (Robinson, 1984; Corbet and Hill, 1991).

Some agriculturalists like to think that all domesticated and cultivated forms are too highly selected to survive in the wild. Some are, but many are not (see section 2.3.2), and that is relevant to the risk assessment of releases of genetically engineered organisms (section 6.5). Domesticated cats, dogs, horses, asses, cattle, water buffalo, sheep, goats and pigs can all be introduced pests. All these forms are sufficiently distinct from the wild ones that Corbet and Clutton-Brock (1984) prefer names other than those used for the wild species. So, they use *Canis familiaris* not *C. lupus, Equus caballus* not *E. ferus, Equus asinus* not *E. africanus, Bos taurus* not *B. primigenius, Bubalus bubalis* not *B. arnee, Ovis aries* not *O. orientalis, Capra hircus* not *C. aegagrus*, and *Sus domesticus* not *S. scrofa*, for the domesticated forms. Cats are often kept specifically to be predators of rodents, so it is not surprising that they are efficient predators of birds and mammals when feral.

As an example of the drastic effect of cats, consider Ascension Island in the central South Atlantic. It is one of the most important warm water seabird stations in the world, with breeding populations of 11 species, four terns (Sterninae), three boobies *Sula*, two tropicbirds *Phaethon*, a frigatebird *Fregata aquila* and a storm-petrel *Oceanodroma castro*, though only the frigatebird is endemic; a shearwater, *Puffinus lherminieri*, is probably extinct there (Ashmole *et al.*, 1994). Because of cats primarily, and ship rats, only the sooty tern *Sterna fuscata* (and a few tropic birds) breed on the main island of about 88 km^2; the others breed on the small Boatswain's Island, about 0.1 km^2, and some sea stacks. Two endemic landbirds are extinct, a rail *Atlantissa*

Figure 5.2 The three invasive species of *Rattus*. Top, common rat *R. norvegicus*; middle, ship rat *R. rattus*; bottom, Polynesian rat *R. exulans*. Ear and tail proportions are useful taxonomic characters. (The top two modified with permission from The Mammal Society vignettes, © Blackwell Scientific Publishers; illustration by Mike Hill.)

elpenor and a night heron *Nycticorax* sp., possibly from direct predation, possibly from the loss of the other nesting birds and other habitat changes. Could the cats (and rats) be eliminated, that would be wonderful for bird conservation (Ashmole *et al.*, 1994).

There are three rats, *Rattus* spp., out of about 50 species in the genus (Corbet and Hill, 1991), that are commensal and which readily become feral. They are: *R. norvegicus*, the common or brown rat (and the laboratory rat), *R. rattus*, the ship or black rat, and *R. exulans*, the polynesian rat. The common rat can be black, many populations of the ship rat are brown, so colour names are best avoided. All three species can be very destructive. For instance, ship rats exterminated the populations of all three endemic species of New Zealand bat (long-tailed *Chalinolobus tuberculatus*, lesser short-tailed *Mystacina tuberculata*, greater short-tailed *M. robusta*) on Big South Cape Island (930 ha) off Stewart Island in about 1965 (Daniel and Baker, 1986). The rats also destroyed five bird species on the island (King, 1984). The greater short-tailed bat is now probably totally extinct, the lesser short-tailed vulnerable or endangered, while the long-tailed is protected but more widespread in New Zealand. The long-tailed is a vespertilionid, a member of a widespread and large family, but the short-tailed species are the only species in the endemic and peculiar Mystacinidae, whose closest relatives are apparently South American (Daniel and Baker, 1986).

On oceanic islands, rats, particularly ship rats, have often but not always been destructive of bird colonies. Atkinson (1985) pointed out that the exceptions are islands that either had endemic rats or land crabs, presumably as a result of selection to avoid predation. The land crabs (*Birgus, Coenobita* and *Gecarcinus*) are widespread in the tropics, the endemic rats were two species on (the Indian Ocean) Christmas Island and several on the Galapagos. All the endemic rats are extinct, except possibly some populations on Galapagos islands as yet without ship rats (Patton and Hafner, 1983; Clark, 1984; Corbet and Hill, 1991).

Grazing can be as destructive of plants as predation is of nesting birds. Rabbits were discussed in Chapter 1. Goats were frequently placed on islands as possible food for ship-wrecked sailors. Carlquist (1965) describes how goats changed Guadelope Island, 250 km off the Pacific coast of Mexico, into a grassy pasture. Some native endemic plants only survived on steep cliffs or offshore islets. Fish too can over-graze. Grass carp *Ctenopharyngodon idella* is troublesome in some parts of the United States (US Congress OTA, 1993). It can destroy habitats for young fish, increase turbidity and diminish waterfowl populations by eating their food. As they can be useful for controlling water weed, the preferred solution now is to use sterile triploids. Grass carp were probably the source of the Asian tapeworm, *Bothriocephalus opsarichthydis*, now found in some indigenous American fish. Welcomme (1992) lists the carp as 'generally viewed as beneficial', suggesting bias in his sample of respondents; it should be at least in the 'mixed feelings' list. Introduced invertebrate herbivores causing severe damage often affect introduced plants, but native vegetation can also be damaged. The gypsy moth is a well-known case.

Examples of introduced predators that have marked ecological effects can be found in all predatory groups, not just in mammals. Another group of undesirable predators are certain ants. Most ants are generalized predators, and can be particularly destructive on oceanic islands (Williams, 1994). They can have other effects too, by farming aphids and mealy bugs. Hawaii has no native ants, but 35–40 species are now found there (Howarth and Medeiros,

1989; Reimer, 1994). Their effect is generally harmful, as the indigenous insects have no protection. The most harmful ants involved include the big-headed ant, *Pheidole megacephala*, the long-legged or crazy ant, *Anoplolepis longipes*, and the Argentine ant, *Linepithema humile* (formerly known as *Iridomyrmex humilis*), which is found up to around 3000 m in Haleakela National Park on Maui. In the Galapagos and elsewhere the little fire ant, *Wassmannia auropunctata*, causes concern for its effect on other invertebrates including ants (Ulloa-Chacon and Cherix, 1994). They also attack reptiles (tortoises and iguanas), snails and tourists. In its natural habitat of central American rainforest, though, it is not a dominant ant (Tennant, 1994), showing once again how difficult it is to predict the ecological effect of a species in a new habitat.

Although ants can be troublesome invaders, the species involved are a very small proportion of the total world ant fauna, a hundred or fewer species out of perhaps 20 000 (Patterson, 1994). Several, such as the Pharaoh ant, *Monomorium pharoensis*, are world-wide pests often described as tramp species.

The last two examples in this section show the difficulty of assessing the impact of predators, for two reasons. First, predators may induce predator–prey cycles, such as may be involved in some of the boom-and-bust examples of section 2.3.3. If so, it may take the system some time to settle down; the initial effect may be much larger than the final one. A possible case involves the freshwater fish *Cichla occelaris*. Second, the ecosystem may be suffering for other reasons when the predator is introduced. It is then difficult to know what changes are caused by the predator and what by pollution, over-fishing, or other insult. The Ctenophore, *Mnemiopsis leidyi*, in the Black Sea illustrates such problems and is discussed after *Cichla*.

Cichla ocellaris (Cichlidae, Figure 5.3), the peacock cichlid or peacock bass, is a South American diurnal predator farmed, among other places, on the Chagres River in Panama. This river runs into Lake Gatun, the largest of the lakes made in building the Panama Canal. The lake is irregularly shaped and

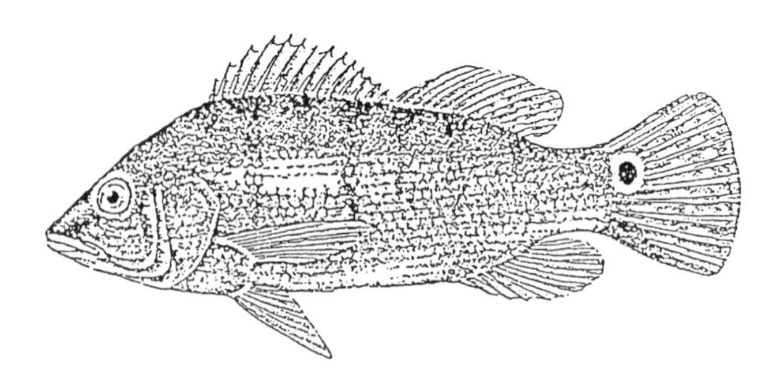

Figure 5.3 Peacock bass or tucanare, *Cichla ocellaris*. This one is about 30 cm long, but individuals can grow to 60 cm. (Copied from Sterba, 1962, with the permission of Urania Verlag.)

large, about 700 km^2 and contains the well-known island and field station of Barro Colorado, as well as 80 km of the canal. The fish was first recorded in the lake in 1969, and had colonized most of it by 1972 (Zaret and Paine, 1973). Initially, although it had no recorded effect on the nocturnal fish species, it had a marked effect on the eight resident species of diurnal fish. Four appeared to become extinct, *Astyanax ruberrimus* (Characinidae), *Aequidens coeruleopunctatus* (another cichlid), *Gambusia nicaraguensis* and *Poecilia mexicana* (both Poecilidae). Another characinid *Roeboides guatamalensis* and *Gobiomorus dormitor* (Eleotridae) both declined by 90%, *Melaniris chagresi* (Antherinidae) by 50%. The eighth, another cichlid *Cichlasoma maculicauda* increased by 50% (Zaret and Paine, 1973).

In 1972, *C. ocellaris* had not invaded the whole lake, so it was a little early to assume that the effects were final. It seems that populations of the species affected survived in the rivers, and made a come-back later, though this has not been properly documented (Welcomme, 1988; Lever, 1994), so the effect of the new predator was less severe than seemed at first. *C. ocellaris* has been introduced in several places and is a species 'whose introduction is viewed with mixed feelings' (Welcomme, 1992). In Florida it is reducing indigenous populations of bass and bream (US Congress OTA, 1993). Sometimes *Cichlasoma* spp. (which are also predators) are introduced as forage for it, and there are 'important local fisheries . . . in Central American lakes' for those, yet none is viewed as beneficial; *C. manguense* is viewed with mixed feelings and *C. bimaculatum* is a pest (Welcomme, 1992).

The last example in this section is a marine one. The Black Sea (Figure 5.4) has suffered severely in recent decades (Mee, 1992). It always was a remarkable sea, entirely land-locked, discharging only through the Bosphorus to the Aegean and Mediterranean, and being fed by several large rivers starting with D, the Danube, Dnestr, Dnepr and Don, which bring it large quantities of nutrients and pollutants. The water is only about 2.2% salt compared with about 3.5% for the open ocean. Most of the Black Sea is very deep, up to 2215 m and averaging 1290 m (Zaitsev, 1992), though there is a wide shelf (less than 200 m deep) in the north, on either side of Crimea, the north-western shelf to the west, the Sea of Azov to the east. The waters deeper than 200 m, about 90% of the water, are anoxic, essentially without oxygen or life, the largest anoxic water mass on Earth. In recent years, because of eutrophication and pollution, the shelf waters are thick with plankton and murkier, and the bottom water there is frequently, though variably hypoxic (severely deficient in oxygen) in summer.

The biological consequences are marked, varied and rapidly changing. There has been mass mortality among the benthos, including a reduction of the rather special ecological community known as Zernov's phyllophora field. This is dominated by the red algae, *Phyllophora nervosa* and *P. brodiaei*, and has shrunk from 10 000 to 50 km^2 (Zaitsev, 1992). There has been a great loss of the associated (and also predominantly red) fauna, including sponges, sea anemones, crustacea, tunicates and fish. Zaitsev (1992) lists, for the whole sea,

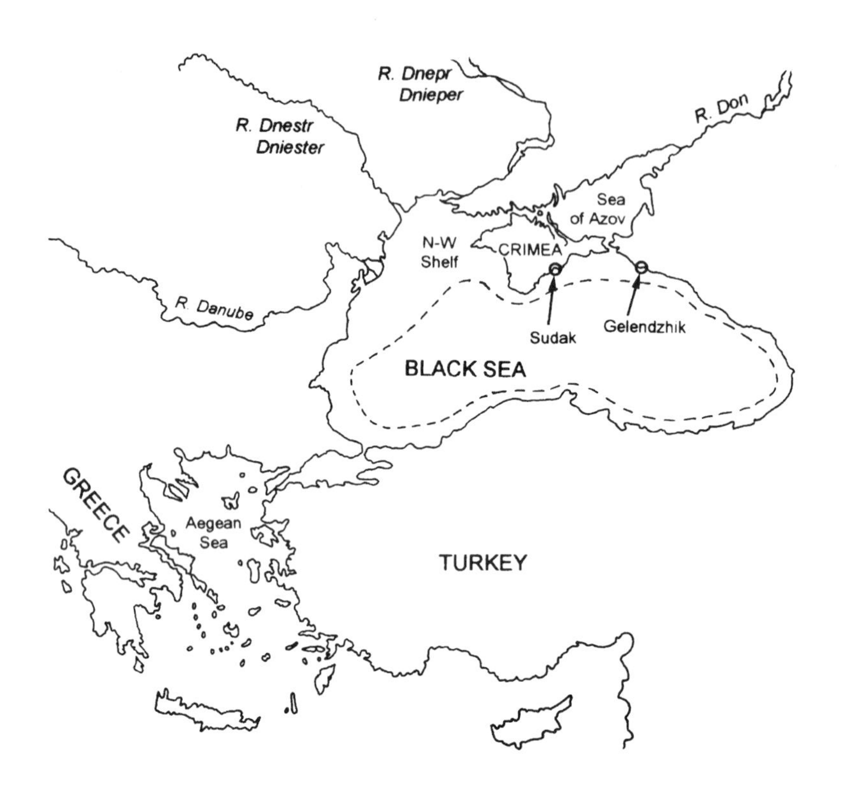

Figure 5.4 Map of the Black Sea and surrounding area, showing places mentioned in the text. (Illustration by Mike Hill.)

26 species that have become very rare or extinct, 14 that have become much more common. The commercial fisheries on the north-west shelf landed commercial quantities of 26 species in the 1960s, but only of five in the 1980s, though over-trawling may be partly responsible. One of the planktonic species that increased spectacularly between the 1960s and the 1980s is the scyphozoan jelly-fish, *Aurelia aurita*, from a population weighing around 1 g m^{-2} (wet weight) of the water column to over 1 kg m^{-2} (Vinogradov *et al.*, 1989; Zaitsev, 1992). Parts of the sea had become a mass of jelly.

Among the introductions that have played a measurable part in the changing ecology of the Black Sea, the most remarkable is *Mnemiopsis leidyi* (Figure 5.5). It is a lobate ctenophore, a comb jelly with two large lobes which, in the adult, are used for catching prey. It is a member of one of the few phyla not found in Carlton and Geller's (1993) ballast samples, but it was no doubt introduced by ballast. A usual size for adults is 20–45 mm, though specimens up to 75 mm are not uncommon (Vinogradov *et al.*, 1989) and lengths up to 120 mm have been recorded (Zaitsev, 1992; Mutlu *et al.*,

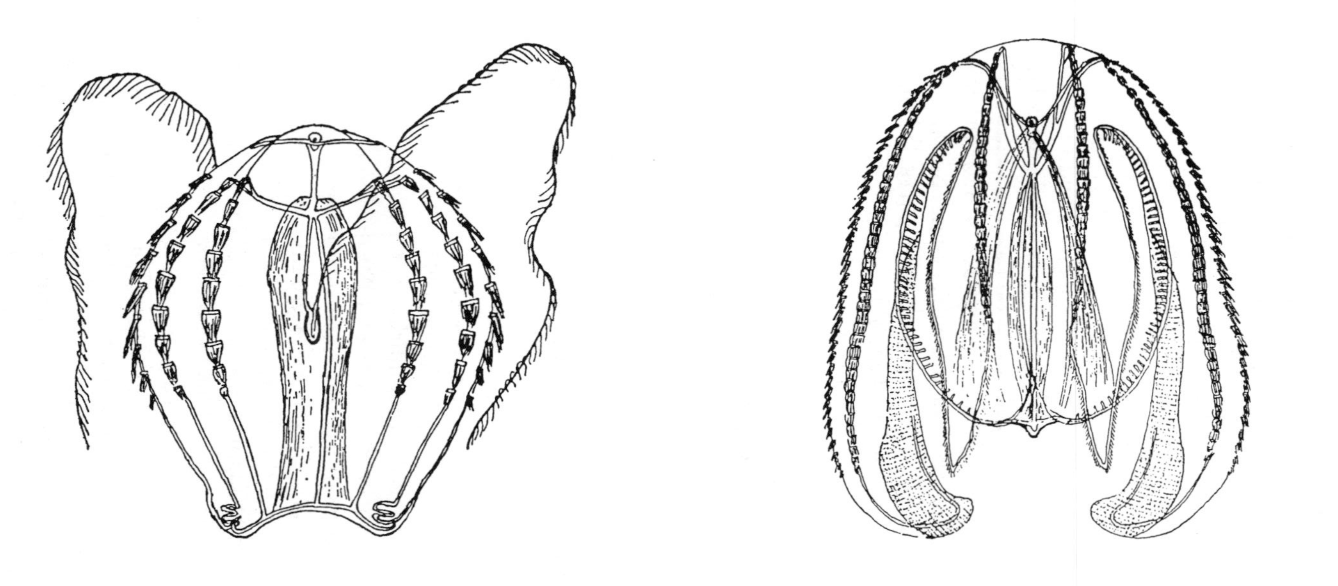

Figure 5.5 *Mnemiopsis leidyi*. (a) An immature specimen showing the typical Ctenophore globular body, mouth at the base, rows of propelling cilia and two tentacles. (b) An adult with the globular body still visible in the centre, but now surrounded by large lobes, and the tentacles gone. Adults can grow to 50 mm across.

1994). An astonishing 300–500 m^{-3} have been seen in the Black Sea (Zaitsev, 1992).

Mnemiopsis is a species of the western Atlantic, particularly estuaries, though it is sometimes found in the open sea. It ranges from Chesapeake Bay, through the Gulf of Mexico and the Caribbean to Venezuela, and recurs off southern Brazil and Argentina (Travis, 1993). The first record in the Black Sea was at Sudak Bay, off Crimea (Figure 5.4) in 1982 but it did not become common until 1987 when large numbers were reported at Gelendzhik Bay, rather further east, and near the Bosphorus (Vinogradov *et al.*, 1989; Mutlu *et al.*, 1994). It is now known in the Aegean and the eastern Mediterranean (between Cyprus and Turkey), and is causing great alarm internationally, largely because of its possible effect on fisheries (Travis, 1993).

How far it depresses Black Sea and Sea of Azov fisheries is difficult to say among all the other ecological troubles there, but *Mnemiopsis* is certainly a most successful invader. In 1988, the wet weight of *Mnemiopsis* increased from 0.4 g m^{-2} in March/April to 1105 g m^{-2} in September, while that of *Aurelia* decreased from 1225 to 24 g m^{-2} (Shushkina and Musayeva, 1990). By that time the biomass of *Mnemiopsis* was an order of magnitude greater than that of all other zooplankton combined (Vinogradov *et al.*, 1989). The sea had changed from one sort of jelly to another. Such a large mass of a predator that can eat animals up to 1 cm long, including young fish, must have a large ecological effect. Such dominance was unknown for any other invader in any other ecosystem.

Since then, the population has decreased, and *Aurelia* has recovered. It seems that *Mnemiopsis* is another boom-and-bust species, though it is still common in the bust phase. A minimum summer population was found in 1991, with about 130 g m^{-2} on average, increasing to around 200 g m^{-2} in both 1992 and 1993, though the distribution was far from uniform in the Black Sea as a whole. *Aurelia* and another native ctenophore *Pleurobrachia pileus* were also both at about 200 g m^{-2} in 1992 and 1993 (Mutlu *et al.*, 1994). It is unlikely, but possible, that this is an equilibrium. *Mnemiopsis* is blamed for the collapse of the Turkish anchovy fishery in 1989, to a quarter of its previous size. Anchovy is a clupeid fish, *Engraulis encrasicolus*, and the fishery remained depressed (at about the levels that were caught in the 1970s, presumably with less effort) in 1990 and 1991 (Kideys, 1994).

It will be interesting to see what happens next. The invasion is certainly a 'dramatic example of the catastrophic effect of ballast water introductions' (Carlton and Geller, 1993), and 'one of the most outstanding global invasion stories in the last 50 years' (Carlton, in Travis, 1993). With the uncertain, and possibly major, effect on Black Sea fisheries, and its invasion of the much saltier (ca. 3.8%) Mediterranean, it is also 'one of the hottest issues that has broken out in the last few years in marine biology' (J. Caddy, in Travis, 1993).

I am left with two gloomy thoughts. The first is that major invasions may come out of the blue at any time. The second is that the state of the Black Sea is even more depressing than most environmental stories, and even less likely to be improved by international agreement in the near future than other cases. I can but hope that the better state and higher salinity of the Mediterranean will obviate a major impact there.

5.3.2 Pathogens

Much less is known about the ecological effects of pathogens than of predators, but it is possible that they are equally important. Their effects are often less conspicuous or harder to prove. Some though are very conspicuous, such as a notable set of pathogens of trees (Gibbs and Wainhouse, 1986; von Broembsen, 1989; Tallis, 1991). Chestnut blight, White Pine blister rust, Dutch Elm disease and *Phytophora* root disease are all well known. I will just sketch the effects. Others such as morbilivirus diseases of seals and lions, which are related to canine distemper, make headlines but their long-term effects are uncertain. The most probable effect is a long-term endemicity, as with myxomatosis in rabbits (section 1.3.2), reducing the host population with consequent ecological effects. In other cases the pathogen burden can be estimated, but the ecological causes and consequences may be obscure.

In another morbilivirus, the effects were striking, though the causal pathway was obscure for a long time. Rinderpest, which is probably the ancestor of measles, swept through southern Africa, killing ungulates, in the 1890s. In some species the mortality may have been 90%. It was thought that wild-life was the reservoir of rinderpest in domestic cattle; vaccination showed the opposite to be the case. Vaccination of domestic herds probably led to the dramatic population increases in wildebeest *Connochaetes taurinus* and buffalo *Synceros caffer* in the Serengeti in the 1960s and 1970s (McCallum and Dobson, 1995). Rinderpest undoubtedly had a major effect on the ecology of many African mammals.

(a) *Tree diseases*

The ecological effects are equally striking in the tree examples. The American chestnut, *Castanea dentata*, was a dominant tree in the species-rich eastern American deciduous forest. It has a large seed about 2 cm long, and it was a notably slow invader after the ice-age (MacDonald, 1993), reaching its climatic limit only about 2000 years ago (Davis, 1983). In some Appalachian forests it made up almost one-third of the large trees, and the forest generally was an oak–chestnut forest, with several species of oak *Quercus* and only one of chestnut. It was removed as a reproducing tree by *Cryphonectria* (previously *Endothia*) *parasitica*, which killed the aerial parts of the trees though leaving the roots alive. The blight spread from New York throughout the forest in the first half of this century (Elton, 1958; Tallis, 1991). Small chestnut trees can still be seen quite commonly, sprouting from the roots, but are almost invariably killed by blight before they can seed.

The effect of the loss of this dominant tree is apparently more economic than ecologic. The wood was very valuable and rot-resistant. Some hill-farmers depended on chestnuts and acorns to maintain their pigs. Acorns alone were not sufficient and much former farm land is now park or forest, as in the Shenandoah National Park in Virginia. The ecological consequences will happen slowly, but seem to involve relatively minor changes in the other tree species. In 50 years an oak–chestnut forest in Virginia (near Mountain Lake Biological Station) became an oak–hickory forest, with three oak species (red *Quercus rubra*, white *Q. alba* and chestnut oak *Q. prinus*) at much their previous densities, while pignut hickory *Carya glabra* had become more common (McCormick and Platt, 1980).

White Pine blister rust has a complex life-history with a stage on species of *Ribes* (currants). It makes growing five-needle pines uneconomic in Britain and is troublesome in North America. It is thought to have come from Asia. *Phytophora cinnamoni* is a root fungus with a very wide host range of over 1000 different species. It has had a particularly severe effect in jarrah, *Eucalyptus marginata*, forests in Western Australia (von Broembsen, 1989). Other species that have been affected included the southern range of *Castanea dentata* before blight struck.

Dutch Elm disease is caused by various strains of the fungus *Ceratocytis ulmi*, and is carried by beetles. It has devastated elms *Ulmus* in both North America and Europe, and again may have come from Asia. An epidemic starting in the 1920s eventually killed 10–20% of the British elms. A later epidemic, starting in the 1960s, which has killed almost all elms in England, seems to have come from North America. The relationships of the various strains are unclear, but the loss of what was a prominent English tree seems to have had negligible effects on other species of plants. Some insects were, no doubt, affected. It was expected that the white-letter hairstreak butterfly, *Satyrium w-album*, would disappear in England, as its caterpillar was said to require elm flowers. In fact, it has survived, living on non-flowering shoots regenerating from stumps. Indeed, more colonies are known than before Dutch elm disease flared up, which shows how a threat can lead to better recording (Emmet and Heath, 1989). It is probably rarer than it was, and its future is uncertain.

Tree diseases are very serious for the species concerned, and for the insects closely associated with them, but seem not to affect other plants to any great extent.

(b) *Animal diseases*

The importance of animal diseases is harder to estimate. Avian malaria has often been given as a reason for the extinction of native Hawaiian birds, the disease being brought in by introduced birds, as was birdpox. The mosquitoes, *Culex quinquefasciatus* and *Aedes albopictus*, that are vectors for the malaria are also introduced. (The spread of the Asian tiger mosquito *A. albopictus* is causing concern in the continental US (Craig, 1993).) Van Riper *et al.* (1986)

found that several native Hawaiian bird species did have malaria, caused by *Plasmodium relictum capistranoae*, but at low incidence, though up to 27% in one mobile species more in contact with introduced birds. Whether the disease is an important cause of extinction, comparable with habitat changes and rats, is far from clear (Pimm, 1991).

Another enigmatic observation is that the introduced starlings and sparrows in North America have a lower parasite burden than the European populations (Dobson and May, 1986). This sort of factor could explain why some invasions fail when others of the same species succeed, and should be remembered when considering the low success rate of invaders.

Diseases may be important in any community. The native British freshwater crayfish, *Austropotamobius pallipes*, has become rare and is now a protected species (Holdich, 1991). One reason is crayfish plague, *Aphanomyces astaci*, a fungus. Although this American disease has been in continental Europe since the 1860s, it is probable but unproved that the latest British epidemic was introduced with signal crayfish, *Pacifastacus leniusculus*, in the 1970s. This species came from California via Swedish farmed stock, certified to be disease-free. It is a very successful farm product and many have escaped. There are several other introduced crayfish now in Britain, and the possible competitive relationships between the species are unclear.

That brings us from vertical (predation, etc.) to horizontal (competition, etc.) food-chain effects.

5.4 COMPETITION, AMENSALISM AND SWAMPING

All these are horizontal food-chain effects, and some examples have already been described. Competition can be thought of as any negative–negative effect between two species, amensalism is a negative–zero effect where one species affects another but is not itself affected. Where there is competitive displacement is often difficult to know which is involved, it is often difficult to measure an effect of the losing species on the winner. So, it is simplest to treat these two together. Swamping, where one species overgrows all others, is more clearly amensalism. Swamping has been seen in *Impatiens glandulifera* (section 1.3.3), *Melaleuca quinquenervia* (section 3.4.6) and *Mimosa pigra* (section 5.1). Competition (or amensalism) was involved in the spread of the grey squirrel (section 4.5).

The squirrel example shows how reluctant biologists can be to accept competition when they cannot see the mechanism. *Cakile*, the sea rocket, is another (Baker, 1986). These are annual plants living near the drift line on the sand and sometimes the shingle of the sea shore. *C. edentula* from the American Atlantic coast reached San Francisco Bay in the 1880s, and spread at an average 65 km per year along the coast from Alaska (Kodiak Island) to the US/Mexico border. In 1935, *C. maritima* was found in Marin County, California, just north of San Francisco Bay. This is a European species; populations from Portugal and the Mediterranean may possibly be sub-specifically distinct from those found throughout north-west Europe (Stace, 1991). Its spread was at about 50 km per year and it displaced *C. edentula* throughout California, and continued south into Baja California. Baker (1986) is puzzled that species of such open communities can be in competition, but that seems to assume that all competition is by physical displacement. With the rapid turnover of annual species, small differences in success at finding suitable sites for germination would lead to competitive displacement. The differences could be in dispersal, fecundity, survival, germination rate, and so on.

A famous example showing that competition is not necessarily scramble or contest is the hares, *Lepus*, in Newfoundland. The native hare is *L. timidus* (or *arcticus*), a circumpolar species (or species complex) ranging from Ireland, Scotland and the Alps across Siberia and the Canadian tundra to Greenland. *L. timidus* is a tundra species, and also a coniferous forest species in the palearctic. On Newfoundland it was found in forest until the snowshoe hare, *L. americanus*, the common North American boreal forest lagomorph, was introduced, for sport, in the 19th century. *L. timidus* is now confined to rough parts of the Newfoundland tundra with boulders, etc. where the hare can take cover from lynx *Felis lynx*. The competitive displacement of the hares seems to have been mediated by lynx predation (Williamson, 1981). If the mechanism is correct, some people would not call the interaction competition at all, others would call it apparent competition. But *Cakile* shows that it is convenient to call all horizontal negative–negative effects competition; it is not, in my view, sensible or convenient to change the name once the dynamics are partially understood. However, the definition of competition is a well-known contentious issue in ecology (Law and Watkinson, 1989).

Competition is often postulated when a new invader becomes conspicuous, but it is now generally realized that population data are needed to show its reality. When the house sparrow *Passer domesticus* reached the American west, it was accused of displacing the house finch *Carpodacus mexicanus*. The house finch was brought for sale to New York in 1940 as the Hollywood finch (Root, 1988) but released when the sales were found illegal, because of the Migratory Bird Act. It has spread up and down the east coast states from New England to Georgia, and has been accused in its turn of displacing house sparrows. Bennett (1990) showed that when the bird populations are examined over a range of scales, evidence for competition either way goes. There are individual interactions, but they seem to have no detectable population consequences. The house finch is the most destructive bird pest in California (Palmer, 1973), so the house sparrow might have won itself some friends if it had reduced the house finch population.

The mechanism of competition, when it does occur, can be subtle and complex, as was seen with squirrels (section 4.5). Petren *et al.* (1993) showed that the competitive dominance of the introduced house gecko (a lizard) *Hemidactylus frenatus* over the native *Lepidodactyus lugubris* is probably largely the result of the small natives moving away from the large invader. The natives are asexual, the invader sexual, which gives the natives a roughly two-fold reproductive advantage. That is insufficient to overcome the disadvantage, from loss of feeding, that comes from avoidance behaviour.

Competition can often be seen more easily in the consequences of successive invasions, than in studying population dynamics at one place and time. Examples include the biological control hymenoptera *Aphytis* (section 3.4.5), squirrels, hares and *Cakile* discussed above, and rats.

Rats apparently compete with each other, though they are predators of mice and young rats too. The ship rat was in Britain (in York and London) in Roman times, 100–400 AD, went extinct in the dark ages 500–700 AD and was established again by Anglo-Scandinavian times ca. 900 AD (O'Connor, 1992). The plague, caused by *Yersinia pestis*, now often called black death (though that is a 19th century invention), is spread in rat fleas, and it used to be thought that ship rats came in with the first major plague epidemic in western Europe of 1348. The archaeological record shows that to be wrong.

Common rats reached the British Isles in the 18th century, and have almost totally displaced ship rats. Ship rats can or could still be found on a few islands such as Lundy in the Bristol Channel, Sark and Alderney in the Channel Islands and Lambay near Dublin (Twigg, 1992; Whilde, 1993), with common rats. Possibly some rocky shore habitats gave the ship rat a refuge; it is the better climber. For the same reason, and as the more tropical species, it used to be found particularly in the roofs of buildings, and near boiler rooms, but with modern handling and control systems it has mostly gone, though there are still a few records from warehouses and docks.

In New Zealand, the polynesian rat was introduced by the Maoris, perhaps around 1000 AD, and was widespread before the arrival of British settlers in the 19th century. It is now found only in Stewart Island (to the south of South Island and the third largest, after that and the North Island), in parts of the fjord country in the south-west of the South Island and on at least 15 small islands (Daniel and Baker, 1986). The ship rat is the widespread New Zealand rat, found all over the three main islands, though only on two of the small islands with polynesian rats. The common rat was apparently widespread in the mid-19th century, but is now very patchy, 'common only in wet areas, on farms, in urban areas and rubbish dumps', and it is on several islands including Stewart Island. It is the sequence of events in both Britain and New Zealand that indicates competitive exclusion, even though tropical islands are known with all seven of the possible combinations of the three species (Atkinson, 1985).

All these examples of competition show how difficult it can be to detect, define and to analyse it. That is perhaps why it has not been implicated often in the serious consequences of invasion. It is all too easy to think there is competition when there is not, or to think there is not when there is. With the other horizontal effect, swamping, these problems disappear. Swamping is a well-known phenomenon. When invading plants have this habit, they can be serious pests. Cases discussed earlier are listed at the head of this section, they come from Britain, Florida and Australia. Here, I will discuss briefly some other land plants from other parts of the world, before dealing in more detail with a floating fern and a freshwater bivalve.

Three other examples are *Prunus serotina*, rum or black cherry, *Helianthus tuberosus*, Jerusalem artichoke, and *Euphorbia esula*, leafy spurge. The first, rum cherry, is a North American species and now a serious pest in the Netherlands (van den Tweel and Eijsackers, 1987). It was introduced in the 17th century, but the present problems come from it being used to afforest heathlands in the 1930s. Its dispersal is rather modest. The second, Jerusalem artichoke, is an example of a crop that has become a pest. Many crops have done that, a point relevant to the risks from genetically engineered crop plants (section 6.5). In Germany, it can form dense stands along water courses, though the species it displaces are often also mostly plants regarded as weeds: *Calystegia sepium* hedge bindweed, *Urtica dioica* stinging nettle, *Galium aparine* cleavers or goose-grass and *Brassica nigra* black mustard (Lohmeyer and Sukopp, 1992). It also occurs in semi-natural vegetation of flood-plain thickets (Kornaś, 1990), and it is naturalized in California and in Catalonia in Spain where they 'have become part of our environment' (Fornell, 1990). It was originally a species of rich and damp thickets in eastern North America. The third one, leafy spurge, is a European plant naturalized over much of North America. It is now 'the worst of the worst' in rangelands in Wyoming and other parts of western USA (Culotta, 1994), forming extensive monospecific stands.

All those are land plants. One of the most spectacular examples of swamping, and an example which illustrates several other points about invasions, biological control and the importance of accurate taxonomy, is a floating fern, *Salvinia molesta* (Figure 5.6). *Salvinia* occurs round the world and has about 12 species, most of them South American. The *S. auriculata* group has a haploid chromosome number of nine, and includes *S. biloba* tetraploid, *S. molesta* pentaploid, *S. auriculata* hexaploid and *S. herzogii* heptaploid. They are used as ornamental plants in indoor aquaria and outdoor fish ponds (Mitchell and Bowmer, 1990).

S. molesta first escaped from a botanic garden in Sri Lanka in 1939, and became a very serious pest over much of Africa, India, south-east Asia and Australasia. It was not properly distinguished from *S. auriculata* until 1972, and its origin in Brazil was discovered in 1978. Although some of the other species of *Salivinia* can be pests (Pieterse and Murphy, 1990), *S. molesta* was far the worst. 'On a world-wide scale the floating weeds are the most important in terms of distress to human life. The "world's worst aquatic weed" is generally agreed to be *Eichornia crassipes* (water hyacinth) which is a free-floating weed; in some parts of the Palaeotropics another floating aquatic *Salvinia molesta* is, today, a more serious threat.' (Cook, 1990). Reproduction is entirely asexual, and one reason that it is difficult to control is that it breaks readily when handled, so it cannot be removed effectively mechanically. It can survive in fragments 1 cm long, grows rapidly and forms dense mats up to 1 m

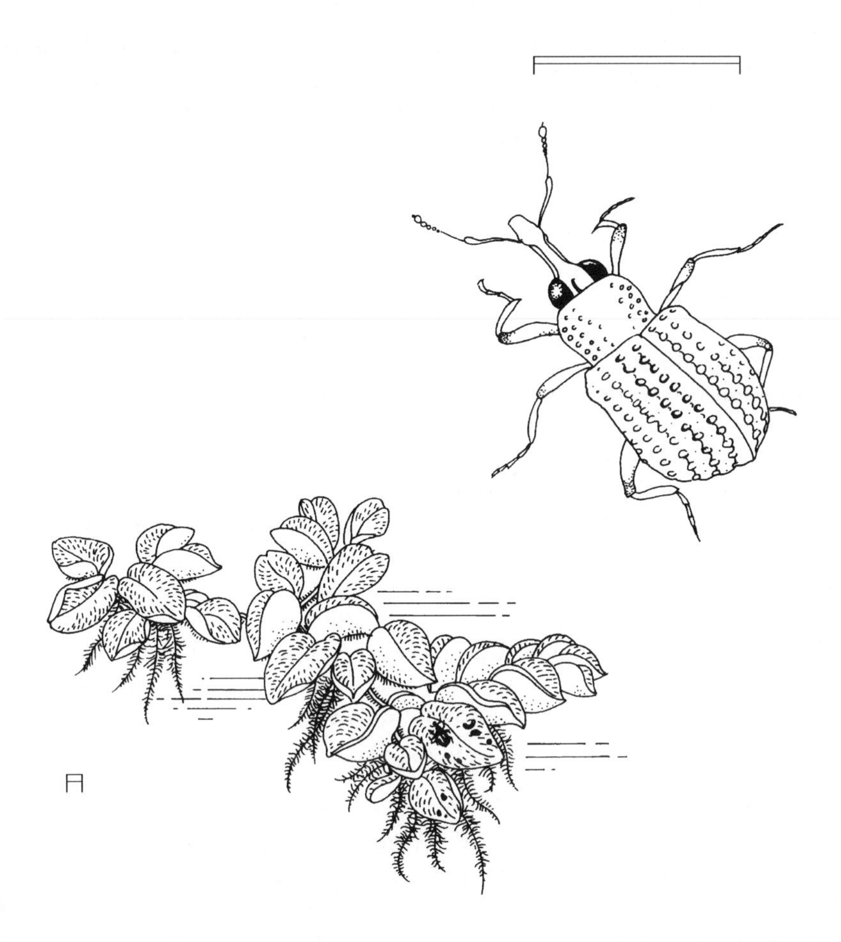

Figure 5.6 *Salvinia molesta* (left) and *Cyrtobagous salviniae* (right). The size bars are each 2 mm. (Illustration by Mike Hill.)

thick. For instance, it was introduced into the Sepik River in Papua New Guinea (see Figure 5.8) in 1971 or 1972, probably from an aquarium. By 1980 it had choked the river, covering 250 km^2, with about 2 million tonnes of plant, disrupting the lives of 80 000 people who relied on the river for transport and food.

Chemical and mechanical control were tried in many places, and proved expensive and ineffective. Three possible biological control agents were found on *S. auriculata* in Guyana and Trinidad in the 1960s: *Cyrtobagous singularis* a curculionid beetle, a weevil, *Samea multiplicalis* a pyralid moth and *Paulinia acuminata* an acridid grasshopper. Although tried in Africa, Sri Lanka and Fiji, none controlled the fern, though all established (Harley and Forno, 1990). A leaf spot fungus, *Myrothecium roridum*, was tried in India (Charudattan, 1990).

In 1980 a new species of *Cyrtobagous*, *C. salviniae* was found in the home of *S. molesta* in south-east Brazil (Figure 5.6). This beetle has been a highly successful control agent, reducing the *Salvinia* populations to a few remnants, except in a few places in Australia which were too cold for it (Room, 1990). In Papua New Guinea, because of the dense growth, the fern was not nutritionally sufficient for the beetle. The solution was to add nitrogen fertilizer! That produced better quality fern that supported the beetle. Feeding by the beetle led to the release of nitrogen which repeated the process. In time the whole area was cleared.

C. salviniae has been used successfully in Namibia and India as well as in Australia and Papua New Guinea, and should control *S. molesta* in most tropical and sub-tropical areas (Harley and Forno, 1990). The moth *Samea multiplicalis* also attacks *S. molesta* in its native home, but is nevertheless an ineffective control, partly because it prefers undamaged plants. *C. salviniae* also attacks other species of *Salvinia*, and is sympatric with *C. singularis* in parts of South America (Room, 1990). The two beetles attack *Salvinia* in different ways. *C. salviniae* feeds mainly on buds and tunnels into rhizomes, while *C. singularis* feeds mainly on exposed tissues, which partly explains their different effectiveness (Sands and Schotz, 1985).

The lessons that it is important to get the taxonomy exactly right when trying to use biological control, that closely related species can give quite different population outcomes, and that only some, if any, of the enemies of a pest will produce significant reductions in the pest population, are all obvious.

The final example of a swamping species is a recent and spectacular invasion of North America, the zebra mussel *Dreissena polymorpha*. It arrived in Lake St Clair, next to Detroit and between Lakes Huron and Erie, in 1985 or 1986, although it was not noticed until 1988. By the end of 1993 the range covered most of eastern North America from Quebec to Minnesota to Oklahoma, Louisiana, West Virginia and the Hudson River in New York state. It forms dense aggregations so that it clogs intakes and distribution pipes, fouls fishing gear, buoys and boats and cooling systems. It smothers unionid clams, and it filters the water so efficiently that the clarity improves, the phytoplankton goes down, and zooplankton and pelagic fish are probably affected (Johnson and

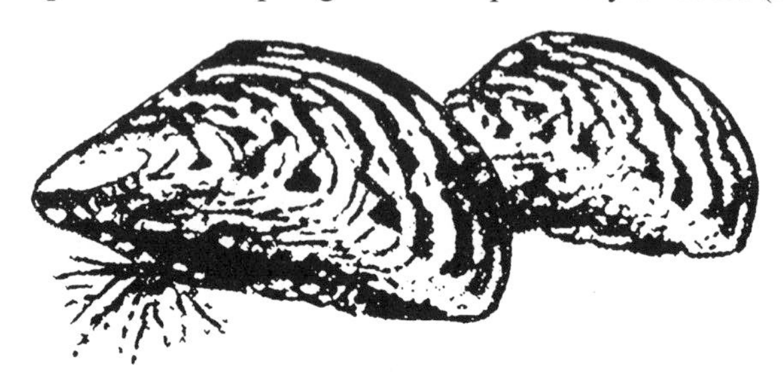

Figure 5.7 Zebra mussel, *Dreissena polymorpha*. Maximum length about 3 cm. (Adapted from US OTA, 1993.)

Padilla, in press). The estimated cumulative loss in the US by 1991 was already $3 372 000 000 (US Congress OTA, 1993).

Apart from the heavy cost, freshwater pests like zebra mussels are particularly difficult to control. There has been no successful biological control of such an organism, and all known potential agents are non-specific (US Congress OTA, 1993), which should make them unacceptable. The suggestion that blue crabs, *Callinectes sapidus*, should be used to control zebra mussels (US Congress OTA, 1993) is an example of the odd ideas put forward to combat new pests. The blue crab is a marine and estuarine species, the basis of a major fishery in Chesapeake Bay. It is unlikely to be able to reproduce in the Great Lakes and the Mississippi River, and it is a general predator. Effective and acceptable biological control agents must be permanent, able to seek out their prey at low densities, and, above all, must be specific.

The zebra mussel apparently originated in the Caspian region in central Asia, and spread across Europe, aided by the extensive canal system then in place, in the 19th century. In Britain it was first found in the Surrey Commercial Docks, London, in about 1824 and had spread through the English and Scottish canal systems by about 1840. It is unusual for a freshwater species in having a planktonic larva, a veliger. It is the only freshwater bivalve in Britain with such a larva, and the only one in which the adult lives on a hard substrate rather than in mud (Macan and Worthington, 1951).

In Britain, the zebra mussel is not a pest. This is partly because of the style of the intake filters used, designed to keep out a variety of organisms that might settle in pipes. In Europe it has occasionally been troublesome in pipes. A fish called the zander or pike-perch, *Stizostedion lucioperca*, was also introduced into western Europe, from further east, in the 19th century for angling. It is now in Britain (Hickley, 1986). In France, the combination of zander and zebra mussel allowed a trematode, *Bucephalus polymorphus*, to complete its life-cycle (Combes and Le Brun, 1990). This led to major mortality in the native small cyprinid (carp family) fish. The sporocyst stage lives in the zebra mussel, the adult in the gut of the zander. The sporocysts produce cercariae which settle on the gills and fins of cyprinids. The zander is infected by eating these fish.

The zebra mussel is but one of several serious invaders that have entered the Great Lakes (Mills *et al.*, 1993). Its concentrations are so great, and its effects so wide-reaching, that it might be said to have an ecosystem effect, the next type of effect to be considered.

5.5 ECOSYSTEM EFFECTS

The distinction between species like *Mimosa pigra* and the zebra mussel, which alter ecosystems in many ways simply by their abundance, and what are generally called ecosystem effects is rather arbitrary. Ecosystem processes are those processes in ecosystems that relate to the system as a whole rather than to individual species. So, water cycling, productivity, cycling of nitrogen and other elements are the major ecosystem processes. Invaders that affect ecosystem processes are those that are said to have ecosystem effects.

Several invasive plants are known to have such effects. The salt cedars *Tamarix ramosissima* and one or more other species with confusing synonyms, often called tamarisk, are Mediterranean invaders of the US south-west. Their deep roots can lower the water table and so drain desert oases, such as Eagle Borax Spring in Death Valley, California. Removal of the tamarisk cured the problem (Vitousek, 1986). In other places there are forests of tamarisks on what had been barren river banks.

Invasive nitrogen fixing plants can have major effects for that reason. Nitrogen fixers are, of course, known to be important in affecting the direction of plant succession (Peet, 1992). Of many possible examples that of *Myrica faya* is perhaps the most dramatic. This is a myrtaceous shrub from the Azores (an archipelago in the North Atlantic), and its nitrogen-fixing symbiont is the actinomycete, *Frankia. Myrica faya* thrives on lava flows in Hawaii. The young lava flows and ash deposits are extremely deficient in nitrogen, and there are no early succession native nitrogen fixers. It not only forms monospecific stands but, by enriching the soil, affects productivity and the structure of the communities that follow (Vitousek, 1986; Vitousek *et al.*, 1987; Vitousek and Walker, 1989).

M. faya is found on sites from volcanic cinder less than 15 years old to closed-canopy rain forest. It is most common in open-canopy seasonal montane rain-forest thinned but not destroyed by ashfall. It is dispersed by birds. In the Hawai'i Volcanoes National Park it increased from 600 ha in 1977 when control was abandoned as too difficult to 12 200 ha in 1985 to 34 365 ha by 1992 (Whiteaker and Gardner, 1992). It added 18 kgN ha^{-1} yr^{-1} in an open-canopy site, a quantity that is likely to have an important effect on the whole system. No suitable biological control agent has been found (Whiteaker and Gardner, 1992). One fungus was discarded for lacking host specificity.

Other plants have direct effects on soil structure. The south African ice plant, *Mesembryanthenum crystallinum*, is an annual which accumulates salt. In California, by so doing, and leaving the salt on the surface when it dies, it suppresses the growth and germination of native species (Simberloff, 1990).

Animals too can affect ecosystem processes fairly directly. Feral pigs *Sus domesticus* have major effects by digging and passing seeds through their gut. This has been documented both for Hawaii and for the Great Smoky Mountains National Park in the southern Appalachians of the USA (Vitousek, 1986; Ramakrishnan and Vitousek, 1989). Guava *Psidium guajava*, strawberry guava *P. cattleianum* and banana poka *Passiflora mollisisma* (a vine) are all plants that benefit from the digging, dispersal and soil fertilization of pigs in Hawaii; the vine in turn has an important effect on mineral cycling. Such effects grade into other severe effects of introductions. How, for instance, would you classify the effect of the common European periwinkle, *Littorina littorea*, an intertidal snail, on the shores of New England? By grazing on the algae on rocks and on the rhizomes of marsh grass, this gastropod has shifted the coastal landscape from mud flats and salt marshes to a rocky shore. By so doing, it inevitably affects many other species (Lubchenco, 1986; Simberloff, 1990). At least these effects mainly alter the balance between species, rather than driving them to extinction, as in the next section.

5.6 EXTINCTION

Extinction is for ever, and so extinction is in one sense the most severe effect that can result from an invasion. That is true for the extinction of a whole species, but local extinctions are also important, and by studying them, it is possible to understand better the processes that lead to global extinction (Williamson, 1989b). So most of the examples in this section deal with islands; some continental extinctions associated with the arrival of *Homo sapiens* will be considered in Chapter 7.

In some cases, the ecosystem effects discussed above are either more serious than local extinction, being the severe disruption of many species rather than the local loss of a few, or lead themselves to local extinction. It is a matter of scale. It can also be a matter of how the species is viewed, as all species are not equal for conservation. The most spectacular and recent global extinction, of smallpox, has been universally welcomed. Where the local extinction is of an introduced pest, that will normally be welcomed too. There can be conflicts of interest, which were discussed in section 5.2.2, but in general an extinction is to be deplored, if only for the loss of biodiversity.

Whether or not a species causes extinctions depends on its biology, not on the way it became introduced. The next two examples are both of extensive extinctions caused by single introduced species. The first, 'the snake that ate Guam' (Pimm, 1987) is an accidental introduction; the second, the predatory snail *Euglandina rosea*, a deliberate biological control one. The case of the snake is well-known, but as it is so recent the accounts differ somewhat (Pimm, 1987, 1991; Savidge, 1987; Engbring and Fritts, 1988; US Congress OTA, 1993). The OTA report gives various figures for the number of species extinct. Here I follow Engbring and Fritts, though the real situation may be even worse by now.

The snake is the brown tree snake *Boiga irregularis* (Plate 5), a skilled climber whose natural range is eastern Indonesia through New Guinea, the Solomon Islands and northern Australia (Fritts, 1988). Those on Guam came from the Admiralty Islands (Figure 5.8) in military traffic (Rodda, Fritts and Conry, 1992). The snakes are largish, up to about 3 m long on Guam and, like almost all Australian snakes, poisonous. They are not fatal for people, though nasty for infants (Fritts *et al.*, 1994). They tend to hide so that they are transported, for instance, in cargo moved by aeroplanes and ships. As a nocturnal animal, fond of hiding in crevices and above the ground, it is easily overlooked.

The snake probably arrived on Guam about 1950 or a little earlier. Guam is slightly less than 50 km long by roughly 10 km wide, 541 km^2 in all, pointing

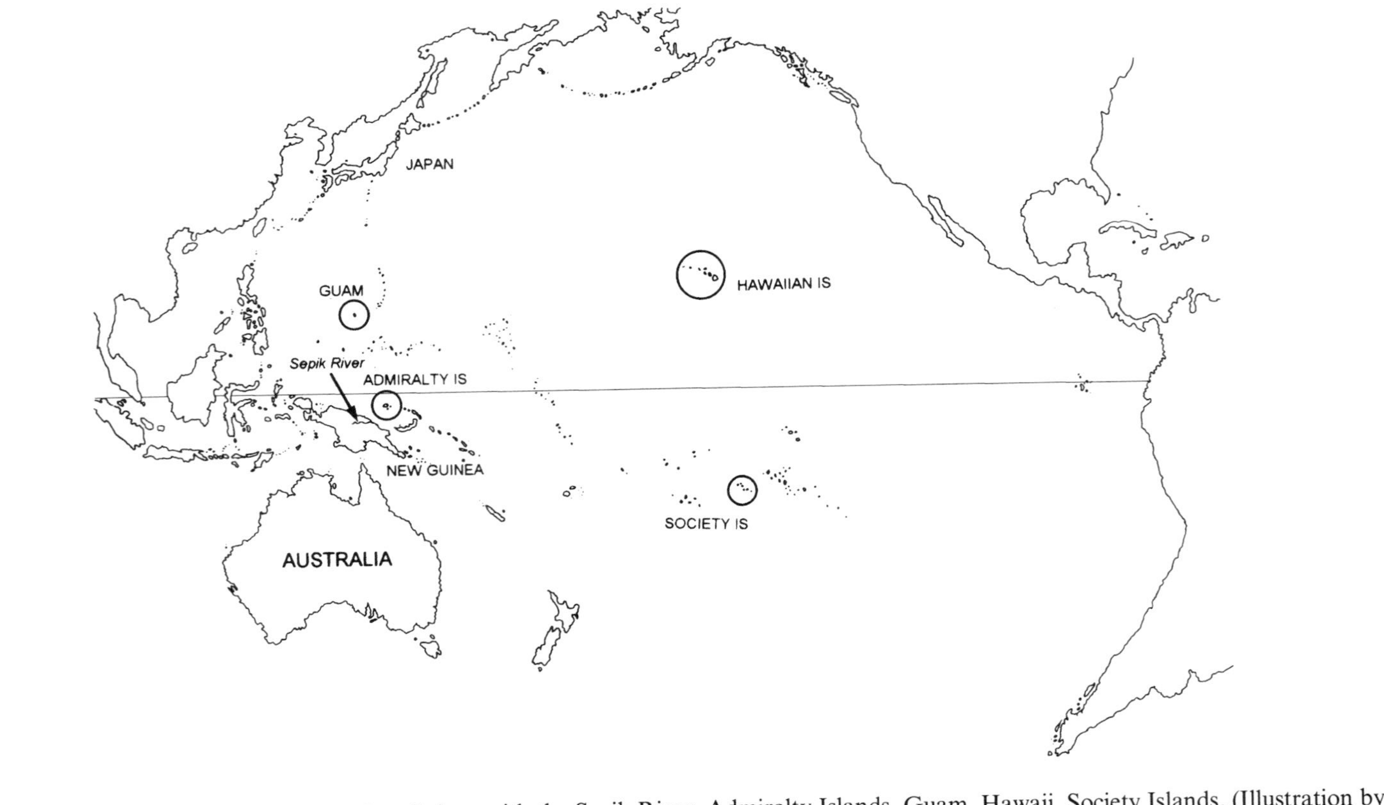

Figure 5.8 Map of the Pacific showing New Guinea with the Sepik River, Admiralty Islands, Guam, Hawaii, Society Islands. (Illustration by Mike Hill.)

roughly north–south. The snake was first common in the south central area, near the sea port, recorded at the tip of the south by 1968, by which time it was well established in the north central area (Savidge, 1987). It was at the north tip in numbers by 1982 (Savidge, 1987), but there may have been isolated individuals there by 1968 (Engbring and Fritts, 1988). The major bout of extinction it caused was in the 1980s, but declines undoubtedly started in the previous decade.

The original vegetation of Guam was largely forest, and the original native birds, 22 species, were birds of forest or wetlands, or seabirds. Development, erosion and wildfires have produced grasslands and there are also urban habitats. Eight introduced species had become established, including the Eurasian tree sparrow, *Passer montanus*. Simplifying by concentrating on the main habitat, there were 12 native and one introduced forest bird, five native wetland birds, four native seabirds, three grassland introduced birds and four introductions either in urban or a mixture of man-made habitats. That leaves one native, *Aerodramus vanikorensis*, the swiftlet that makes the nests of birds' nest soup, and which nests in caves. (The taxonomy and English names of swiftlets are both in flux at present (Pratt, Bruner and Berrett, 1987).) Nearly every bird that nests where snakes can reach them, essentially all tree nesters and some others, have either become extinct or very rare. Some were extinct before the snake arrived. Guam has been described as an avian desert.

There were four early extinctions, before the snake arrived, one seabird (a shearwater), one forest bird (a megapode) and two aquatic birds, a rail (*Porzana*) and a duck. As with other islands, habitat changes and rats are likely to have been the main causes. A reed-warbler, *Acrocephalus luciana*, was extinct by 1970, possibly because of the snake. Since then nine of the other eleven native forest birds have become extinct and two rare. The rare birds are first a crow, *Corvus kubaryi*, which is failing to raise young, perhaps because of predation and nest disturbance by snakes. Only old birds are left. The second is a starling, *Aplonis opaca*, which is however doing well on the small subsidiary island of Cocos, of a few km^2. The one introduced forest bird is rare, but as it is the *Gallus gallus*, the red jungle fowl or feral domestic chicken, its status is, as usual, doubtful. The three remaining seabirds are also all rare, probably largely because of nest predation by the snake. The only native birds that are not rare are a bittern and a reef-heron with nests in habitats the snake does not like, and which may be able to defend their nests. None of the introduced birds is extinct, but the snake is not common in grassland and urban habitats. They are scarcer than would be expected without snake predation.

There are captive populations of two species recently extinct in the wild, the Guam rail, *Rallus owstoni* and the Micronesian kingfisher, *Halcyon cinnamomina*. These could be re-introduced were the snake ever to be eliminated, but there is no known way of doing that. However, goats have been removed from some relatively large Galapagos islands (Hoeck, 1984), showing that the apparently impossible can be done, though of course it is easier the smaller the island.

The devastation caused by one species on one island is remarkable. The ability of the snake to stow away in cargo is worrying, particularly as Guam is an aerial cross-roads and a major American military centre. The snake has been found on arrival at Hawaii several times (US Congress OTA, 1993), and on other Pacific and Indian Ocean islands. It seems only a matter of time before some other island avifauna is destroyed. The snake appears to survive by eating lizards. Even ship rats, which were common, and introduced shrews *Sonchus murinus* are now rare in Guam forests. From the rather thin data, it would seem that the brown tree snake takes 15 years or longer to eliminate almost all bird species in forest. The ecological changes on Guam include the loss of vertebrate insectivores, pollinators, seed dispersers and a frugivore. Subsequent changes in plants, arthropods and their predators are evident (T.H. Fritts, personal communication).

The next case is equally horrific. The Pacific islands had a mass of interesting and beautiful tree snails, Achatinellids and Partulids. The evolution of *Partula* on Moorea in the Society Islands, the next island to Tahiti, is a classic example of allopatric and adaptive speciation (Johnson, Murray and Clarke, 1993). The details of the evolution of the nine partulid species there were worked out from 1962 by Clarke and Murray in a highly regarded series of studies. There are, or were, 57 partulids in the Society Islands, though only between two and eight species in other archipelagos such as Marquesas, Australs, Cooks, Samoa, Vanuatu, Solomon Islands and the Marianas (including Guam) (Cowie, 1992). Hawaii had perhaps a hundred species of *Achatinella*, though the taxonomy is in doubt (Carlquist, 1980).

The giant African snail, *Achatina fulica*, has been introduced to most Pacific islands, and is a pest in crops and gardens. To control it, there has been the exceptionally foolish introduction of a predatory snail, *Euglandina rosea*, a native of Florida and Central America (Plate 6). *Euglandina* has eliminated indigenous species even faster than the brown tree snake. On Moorea, it eliminated all partulids in 10 years. It will probably eliminate all endemic tree snails throughout the Pacific. In the Society Islands, it was on all islands apart from Huahine (Cowie, 1992) but it is there now (B.C. Clarke, personal communication, BBC *Horizon*, The predator, November, 1994). On Hawaii, the Achatinellids were already much reduced by habitat changes, rats and probably ants, but *Euglandina* is there now and implicated in some extinctions.

It is not even at all clear if *Euglandina* has an appreciable effect on *Achatina* populations, though it does eat giant snails as well as small ones. The Moorean partulids are being maintained in captivity, partly in the hope that *Euglandina* will die out now that it has destroyed all tree snails. Regrettably it seems to be surviving at a density of one to every 10–20 m^2, possibly on forest litter snails (B.C. Clarke, personal communication). Conservation is full of sad stories, but there can be few sadder than this, the totally unnecessary and pointless destruction of fascinating biodiversity by a deliberate introduction. As it was all in an attempt at biocontrol, it is the sharpest reminder that biocontrol must be absolutely specific (and remain so) to be safe.

None of these introductions would have happened without human help, intentional or not. Wherever people have colonized isolated islands, extinctions have followed. Some come from hunting, some from habitat change, for agriculture or other reasons, some from introduced species such as rats. Milberg and Tyberg (1993) list both more than 200 species of bird only known as sub-fossils and also another 160 cases in which extant species are only found sub-fossil on particular islands. Those are all pre-historic extinctions (see also section 7.4.3), and all but a few are believed to be the result of human colonization in one way or another. For birds in historic times the three major causes both of extinction, and of reducing populations so that they are endangered, are habitat loss, predation and hunting. Habitat loss is involved with 19% of extinct island birds, and 58% of those endangered. Hunting has affected 15% and 26%. Predation, which is mainly predation by introduced invaders, is a factor in 42% of extinctions, and for 40% of those endangered (King, 1985). Some 93 species and 83 subspecies, at least, of island birds have become extinct since 1600. Extinction by introduced species is not that rare an event.

5.7 CONCLUSION

When invaders have important effects they can be very serious indeed. The examples in this chapter show the great variety, as well as the great cost, of a small selection of invaders. Even though the chances of an invader establishing, and the chances of an established species becoming a pest, are small (the tens rule), the costs are so large that it is sensible to take steps to minimize the number of introductions. Many countries of course do, with quarantine.

Quarantine is, at best, only moderately effective. Removing, or reducing to an acceptable density, an unwelcome invader will normally require control. Chemical control is costly, and has to be kept up, quite apart from any undesirable side effects it may have. Biological control, when it works, is permanent and usually cost-free after release, though the earlier stages are expensive. In practice in most cases, a mixture of methods, including habitat management, will be necessary (Groves, 1989). The examples in this chapter show how important it is that biological control should be absolutely specific. Even then, there are evolutionary worries, which I will consider in the next chapter.

The effects discussed in this chapter have been ecological ones. There are also genetic ones. The next chapter discusses the importance of genetic factors both as causes and as effects of invasions.

6 Genetic and evolutionary effects

Conceptual framework points (from Table 1.1):

CFP 8 Genetic factors may determine whether a species can invade; genetic factors affect events at the initial invasion; evolution may occur after invasion

CFP 10 Invasion studies are relevant to considering the risks of introducing new species or genotypes, the release of genetically engineered organisms and the success and consequences of biological control

6.1 INTRODUCTION

From the ecological consequences of invasions, I now turn to the genetic aspects. Genetic differences are important not only in the effect of invasions, but are also part of the reason for the success of some invasions and the failure of most. This chapter is concerned with all aspects of the interactions of genetics and invasion biology. The subject is full of promise but unsatisfactory, because, until now, it has usually been impossible to identify the individual genes involved either in favouring invasion or in changing after an invasion. Molecular ecology may improve that (Williamson, 1992), but it will take some time. The possible risks from the release of genetically engineered organisms (section 6.5) gives urgency to such research.

There were some examples of the relevance of genetics to invasions in the first chapter. First, genes may affect the success of invaders. In *Impatiens* (section 1.3.3) two allospecies, two species largely allopatric but partly overlapping in range, *I. capensis* and *I. noli-tangere*, behave ecologically very differently in England. The latter is a rare and declining native now confined to the north-west, the former is a successful North American invader spreading in, but apparently limited to, the south and east. The characters separating them are few and apparently trivial; it is reasonable to assume that a very small proportion of the genome is involved. The genetics of species differences are discussed in the next section, but the differences in ecology clearly have a genetic basis. The second point from the introductory chapter is that gene frequencies may change after invasion. Rabbits adapted to the myxoma virus, and vice-versa (section 1.3.2). Again, the genes involved have not been identified, though the effect is clearly genetic.

The third example in the introductory chapter was the fulmar (section 1.3.1), where it is possible that a genetic change or changes affecting nesting behaviour might have been important in the spread across Iceland and beyond. No gene is known, but in this case the spreading populations can be compared with the colonies at St Kilda by-passed by the invasion, as well as with the arctic colonies. Better evidence is needed, and that is a common problem with the genetics of invasions.

Perhaps because genes could not be identified, there has sometimes been a tendency to postulate genetic changes when there are unexpected ecological changes. For instance Mayr (1965) argued that 'The rapid colonization of temperate Europe by Serin Finch [*S. serinus*] and Collared Dove, thus, seems to have had genetic reasons in addition to the external ones', because the spread was too fast to be a response to climate change. He allows that 'In both cases it seems that the beginning of the northward movement was favoured by an amelioration of the climate and the spread of the favourite habitat of these birds'. Mayr may be right or wrong about genetics in those birds; I hope someone will soon find out. But in both cases it is perfectly possible that the habitat changes alone were responsible, provided that the birds were just responding to certain key aspects of the habitat rather than to all aspects.

Similarly, entomologists have suggested that genetic changes are involved in the establishment and in outbreaks of introduced insects, both involving an increase of population. Neither Myers (1987) nor Mitter and Schneider (1987) found much support for this suggestion for outbreaks. The latter conclude 'the subject of genetic effects on numerical change consists largely of ill-defined hypotheses and scraps of inconclusive evidence'.

So it is well to be cautious when genetic effects are postulated. Nevertheless, there are cases where genes are important as the *Impatiens* and rabbit–myxoma examples show. Genes may affect whether a species can invade; there may be genetic changes contemporaneous with the invasion that are important. In the long run, there will be genetic change after many invasions, but short-term effects can be important too. This chapter deals with all these aspects before uniting them in a consideration of the risks from the fast-developing topic of the release of genetically engineered organisms. Of particular importance is the fact that a small genetic change can have a large ecological effect.

6.2 GENETIC DIFFERENCES ALLOWING INVASION

The major evidence for the importance of genes on the success of invaders is the contrasts between closely related species, such as *Impatiens* (section 1.3.3), sparrows (section 2.4.3), and rats (section 5.4). Before considering one more case, and what is known of the genetics of species differences, there are a few instances where intra-specific genetic variation is important. Two major cases are influenza (section 6.2.1) and the Africanized bee in the Americas (section 6.2.2), but there are some minor ones too.

It has often been difficult to determine whether genetic factors are involved. The spread of the starling in North America started in New York, but there were unsuccessful introductions in Quebec and elsewhere (Long, 1981). Were these failures from the wrong genes, from too many parasites (section 2.4.2), from too small a release, or just from bad luck with predators, the habitat or the weather? Two contrasting cases show that intra-specific genetic variation can be important in the field (Williamson, 1992).

Rhizobium is the bacterium that forms root nodules on leguminaceous plants and fixes nitrogen. *R. trifolii* forms nodules on subterranean clover *Trifolium subterraneum*. Usually it can also nodulate white clover *T. repens* but not peas *Pisum sativum*. A single gene difference prevents nodulation on *T. repens* but, surprisingly, allows it on *P. sativum* (Drodjevic *et al.*, 1985; Young and Johnston, 1989). That is an example of a gene difference affecting the habitat (in this case a plant root) that can be invaded.

The other field example is of an invasion of a pathogen being limited by the genotype of the host. Skeleton weed *Chondrilla juncea* is a Mediterranean plant that is a serious weed in Australia and elsewhere. Sometimes it can be controlled by a rust fungus, *Puccinea chondrillina*. Skeleton weed is a triploid obligate apomict, and three genetically and morphologically distinct races have been distinguished. *P. chondrillina*, or at least the strain that was introduced, only attacks one of these races (Burdon *et al.*, 1981). Apomicts are plants that produce seed without fertilization.

There are also two laboratory studies, again microbial and pathogenic, which show the importance of single genes. For *Yersinia pseudotuberculosis*, invasion of the cells of its host is central to its pathogenicity. Isberg and Falkow (1985) isolated a gene responsible, which they called *inv*, and inserted it into the standard laboratory bacterium *Escherichia coli* K-12. K-12 is not a pathogen, though some other strains of *E. coli* are. The *Yersinia* gene made K-12 into an invader of cells in culture. In the other study, Schäfer *et al.* (1989) working with fungi took the *pda* gene from *Nectria haematococca*, a pathogen of peas *Pisum sativum*, and put it in *Cochliobolus heterostrophus*, a pathogen of maize or corn *Zea mays*, which then also became virulent on peas. The enzyme produced by *pda* is pisatin demethylase, a cytochrome P-450 monooxygenase that detoxifies the phytoalexin pisatin. That is why *pda* allows *Cochliobolus* to invade peas. With that action, it is not surprising that *pda* failed to change a saprophyte, *Aspergillus nidulans*, into a pathogen.

All these examples showing single-gene effects are cases where another organism is the habitat. I know of no cases where a single gene has been shown to be important in other habitats, but small sets of genes can be.

An indication that a few genes may be involved in successful invasions comes from the collared dove, *Streptopelia decaocto*, though not in the way that Mayr discussed. He was trying to explain why it started to spread at all

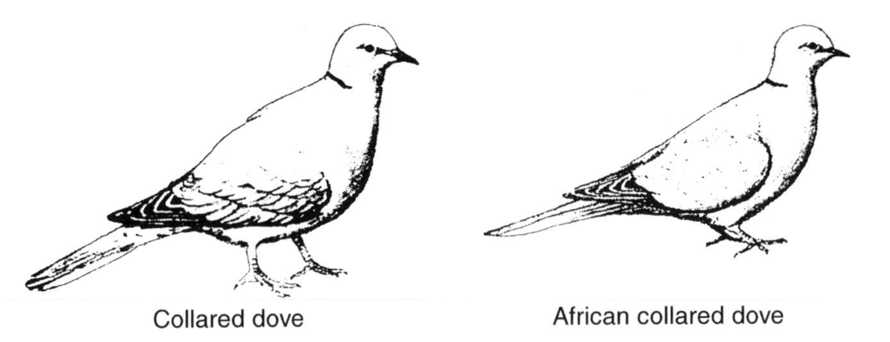

Figure 6.1 Collared dove, *Streptopelia decaocto* and African collared dove, *S. roseogrisea* (the ancestor of *S. risoria* the barbary dove, a cage bird). (From Harrison, 1982, with permission from Harper Collins Publishers.)

in Europe. *Streptopelia decaocto* is one of a group of allopatric species, a super-species. The others are *S. roseogrisea* from the Sahel, Africa just south of the Sahara desert, and (probably) *S. reichenowi* from a little further south, around the Giuba valley in Somalia (Cramp *et al.*, 1985), and *S. bitorquata* from Java and the Philippines (Goodwin, 1970). There are other Asiatic and African *Streptopelia* species; most can easily be crossed in captivity, so Snow (1978) is cautious about how they should be grouped. There is also a cage-bird, the Barbary dove, called *risoria* by Linnaeus. All those three authorities say it is a domesticated form of *S. roseogrisea* (Figure 6.1).

Collared doves are as commensal as house sparrows in Britain (Lack, 1992) and many spend their winters on farms feeding on agricultural and other seeds. Barbary doves, *S. risoria*, have bred on a number of occasions in England (Fitter, 1958; Sharrock, 1976) but never established. It seems they find the British winter too difficult, which is consistent with a derivation from *S. roseogrisea*. Collared doves spread across Europe from Turkey where the winters can be cold. So, the reason why Barbary doves have not established, but collared doves have, may be to do with those genes responsible for winter tolerance in the collared dove. Barbary doves have established populations in some southern cities in the USA (Los Angeles in California, Tampa in Florida, probably Tucson in Arizona and possibly Panama City in the Florida panhandle (Root, 1988)), all with warm winters.

So how many genes do distinguish closely related species? This is an old question, and part of the answer comes from the early days of the evolutionary synthesis. The genes that distinguish species are much the same as the genes that distinguish individuals in a species (Orr and Coyne, 1992), and there are not many more of them. The genetic distance between sibling species is only slightly greater than the genetic distance between conspecifics. That still does not say how many genes are involved, but does suggest that the number of new genes required for speciation is not large.

For morphological differences in plants, Gottlieb (1984) concluded that discrete differences in structure, shape or architecture are often controlled by only one or two genes but that continuous variation, such as differences in height or yield, is polygenic. Polygenic inheritance is when several genes have roughly equal effects on the phenotype; when one gene has the predominant effect it is called a major gene. The distinction is essentially a phenotypic one, and distantly related to the primary gene action. Polygenic effects can be pleiotropic effects of major genes, though as Gottlieb (1984) points out, genes acting early in development are more likely to show simple patterns of inheritance, to be called major genes. Similarly, in animals, qualitative differences in pattern or coloration are often controlled by major genes, quantitative characters polygenically (five or more genes) (Maynard Smith, 1983).

The relative importance of polygenes and major genes in species differences is still obscure. Maynard Smith (1983) says, 'data on species hybrids lead to the conclusion that some differences are polygenic . . . and others involve major genes. . . . This may be a disappointing conclusion to those who like clear-cut answers'. But even when there are polygenic effects, the number of genes is, in the few cases analysed, less than ten. For adaptive differences in characters separating species or subspecies, Orr and Coyne (1992) found three cases of one gene responsible, eight of two or more, one of five and one of six. Tauber and Tauber (1987), studying variation in the seasonal behaviour of lacewing insects (*Chrysoperla*, Neuroptera), found 'The variation in these response patterns underlies important ecological differences and, in some situations, reproductive isolating mechanisms. Secondly, very few Mendelian units account for the ecologically divergent and reproductively isolated populations'.

From hybrid zones, Barton and Hewitt (1985) give estimates for the total genetic difference, between the taxa meeting and hybridizing, of a few hundreds (150 in *Podisma*, a grasshopper, 500 in *Bombina*, a toad). They think the differences result from 'slow evolution, relative to the rate at which population restructuring brings together independent tension zones' rather than from some sort of genetic revolution or pervasive co-adaptation. If there are 10 to 20 genes for each of 10 to 20 characteristics involved in species differences, this estimate is of the same magnitude as those given above.

For invasions, the conclusion from this handful of studies is that the critical difference between success and failure will often come from differences at around 10 genes or fewer. That is consistent with the examples above. But bearing in mind the considerable genetic variation within populations, most allelic changes will not affect the potential for invasion. There are two cases of invasion where rather more is known of the importance of genes: influenza and the Africanized honey bee in the Americas.

6.2.1 Influenza

Influenza is a disease found in several species of birds and mammals and is caused by a single-stranded enveloped RNA virus (Webster, 1994). The vertebrate body fights it with the immune system, developing antibodies. New antigens, ones not recognized as familiar by the immune system, can cause a more severe disease. There are in fact three viruses, influenza A, B and C. This account is restricted to influenza A, the most important of the three.

Serious outbreaks of influenza are usually caused by new genetic variants of the virus, even though, as Kilbourne (1980) says 'Paradoxically, the disease itself has remained a stable and recognizable entity', clearly identifiable, for instance, from ancient descriptions. At irregular intervals, a few times in each century, pandemics have swept across the world. Table 6.1 gives Patterson's (1986) list of pandemics and probable pandemics since 1700. His maps show that the disease usually spreads at something like 100 km a month. For the earlier pandemics, the outbreaks in Europe were all first recorded in Russia, but could well have started further afield. The 1918 pandemic is well known for having killed more people than the battles of World War I; deaths from flu were probably between 20 and 40 million (Webster, 1994).

The right-hand column of Table 6.1 gives the viral type, indicating the major genetic shifts known to be involved in the later pandemics. The types of the earlier outbreaks are not known, though it is possible that with recent advances in techniques for studying ancient nucleic acids they could be characterized at some time. Between pandemics there are genetic drifts, meaning minor changes in the viral genome, not the population process of genetic drift. Distinguishing drift from shift in historic records is difficult. Patterson does not include the epidemic of 1743, when the disease got its present name (Pereira, 1980), as a pandemic. Influenza is the Italian word corresponding to the English word influence, but in Italian it has a rather different range of meaning, including epidemic.

Table 6.1 Pandemics and probable pandemics of influenza since 1700*

Year	*First report*	*Viral type*
1729–1730	Russia?	
1732–1733	Russia	
1781–1782	Russia, China?	
1788–1789	Russia	
1830–1831	Russia, China	
1833	Russia	
1836–1837	Russia?	
1889–1890	Russia	H2
1899–1900	?	H3
1918	France, USA	H1N1
1957	China	H2N2
1968	China	H3N2
1977	China	H1N1

*Simplified from Patterson (1986).

The genome of influenza A has eight separate segments, which come as eight rings of ribonucleoprotein in the virion or virus particle. The complete sequence is known for many different strains (Palese and Garciá-Sastre, 1994). Most segments code for just one major protein, of which there are seven. For epidemiology the two most important are haemagglutinin (or hemagglutinin, *HA*, segment 4) and neuraminidase (*NA*, segment 6). In the virion these are in glycoprotein spikes sticking out from the lipid-covered envelope that encloses the eight ribonucleoproteins, and so they are important antigens. *HA* attaches the virus to cells (red blood cells in the original observations); *NA* attacks a receptor and its main function is probably to promote the release of budding virus particles from the host membrane (Palese and Garciá-Sastre, 1994; Webster, 1994).

Antigenic drifts come from mutations in the *HA* and *NA* genes, and usually produce a minor loss of immunity in the population and so at most a minor epidemic. Antigenic shifts probably usually come from recombination, and may often involve a shift of the virus between different hosts. Swine flu is a well-known disease, but influenza occurs in many other species, horses, cattle, dogs, cats, monkeys, baboons, gibbons, chickens, geese, ducks, turkeys, terns and shearwaters (Kilbourne, 1987). Sometimes the disease in these species is like that in humans, sometimes there is a species-specific disease, sometimes no symptoms at all. The role of these species in genetic shifts is still controversial. What is not in doubt is that pandemics start with a genetic shift, as the right hand column of Table 6.1 shows. The various H and N numbers refer to major variants of the *HA* and *NA* genes.

Pandemics in influenza are invasions, if rather unusual ones. The new territory is the set of people not immune to the new virus type. The pandemic advances in the usual wave. What influenza shows more clearly and precisely than any other example is that the shift in a single gene is sufficient to allow a new successful invasion.

6.2.2 Africanized bees in the Americas

The spread of a new form of honey bee *Apis mellifera* (Plate 7) from Brazil across South America and up through central America (Figure 6.2) is a spectacular invasion with economic and other consequences for human life. Precisely what has happened has been disputed, as has been the right name for these bees. One thing is sure, the bees are genetically different from those that were there before.

European bees can be dangerous, particularly to bee-keepers who have been sensitized by previous stings. There are 4–5 deaths from insect stings each year in the UK (40–50 in the USA), mostly of bee-keepers. Africanized bees have been dubbed killer bees and are much more dangerous. They cannot safely be kept near people or livestock or handled by amateurs. Africanized bees can attack with no provocation and inflict hundreds or thousands of stings. There have indeed been many deaths. Winston (1992) describes the

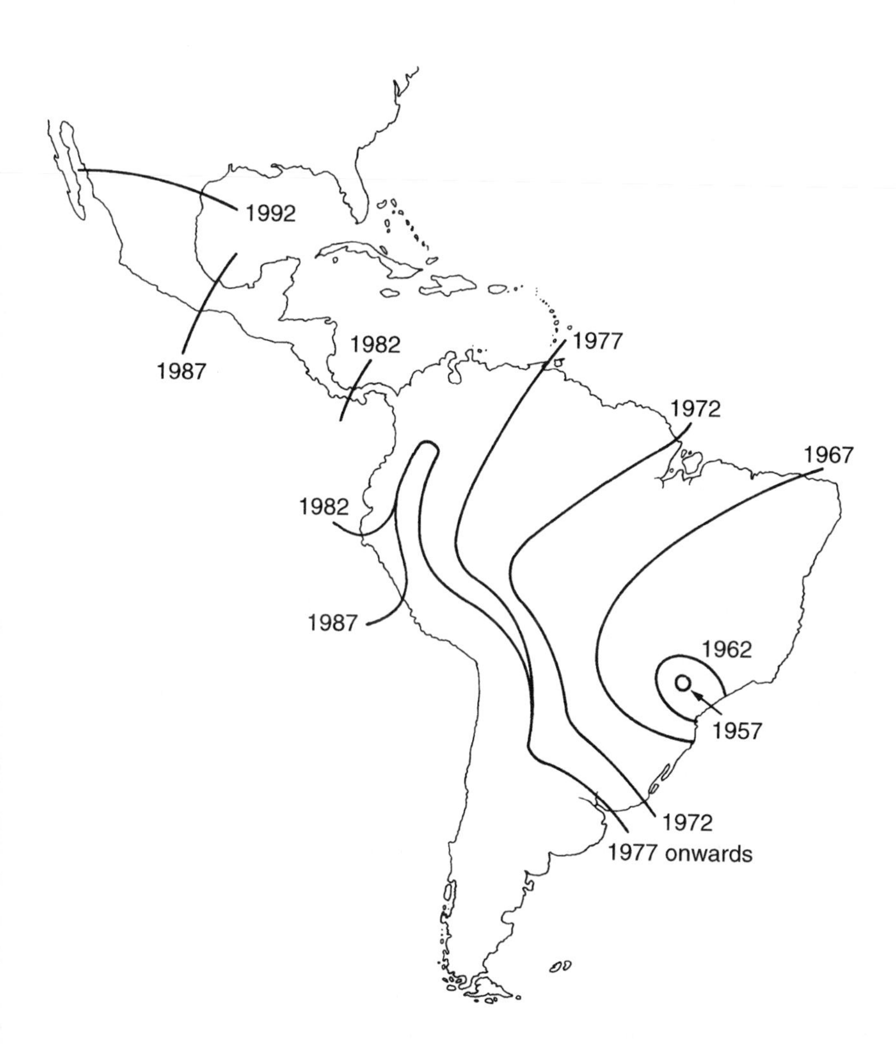

Figure 6.2 Map of the spread of the Africanized honey bee, *Apis mellifera*. The contours are at 5-year intervals, and are in places interpolated. (Compiled from several sources; illustration by Mike Hill.)

problems with Africanized bees and the measures that can be taken to minimize the effects.

The original centre of *Apis mellifera* was probably in the Middle East, possibly in or near Iran (Garnery, Cornuet and Solignac, 1992). Its closest relative is *Apis cerana* of India; hybrids die at the blastula stage. There are also two giant and one dwarf species of *Apis* in India and further east (Winston, 1987). *A. mellifera* has been somewhat domesticated, and moved to new localities, for thousands of years. Bee taxonomists recognize many subspecies (Winston, 1987; Smith, 1991) but here we need only distinguish temperate ones from western Europe and tropical ones from Africa. The former include *A. m. mellifera* of north-west Europe and the Iberian peninsula, *ligustica* from Italy and *carnica* from eastern Europe. All these and their hybrids are used for bee-keeping in the Americas. The only tropical race to be mentioned is *scutellata* which ranges from East Africa south to the Transvaal.

Female bees are either queens or workers, the difference depending on the feeding of the larvae, and are diploid. Males, drones, are haploid. So the genetics is somewhat unusual, with all chromosomal genes sex-linked. Mitochondrial genes have the normal maternal inheritance, descending through eggs only, not sperm. Although bees have not been much changed genetically by domestication, it is possible to select for say colour, yellower or blacker, and for some aspects of behaviour.

Because bees depend on flowers, temperate races have to be adapted to a cold winter when they cannot forage. Tropical bees can normally forage throughout the year. That difference explains most of the suite of characters associated with the two types (Table 6.2, which is an abbreviated summary of a complex situation). Temperate bees need large protected nests with lots of honey to see them through the winter. Colonies reproduce by swarming, when a queen with workers leaves the hive to found another, leaving another queen

Table 6.2 Characteristics of typical temperate and tropical bees. These are tendencies; there is much variation and overlap in the characters*

	Temperate	*Tropical*
Colony size	Larger	Smaller
Honey storage	Larger	Smaller
Exposed nests	Rarer	Commoner
Swarming rate	Lower	Higher
Absconding	Rarer	Commoner
Colony defence	Moderate	Intense
Development period	Longer	Shorter
Worker adult life span	Longer	Shorter
Foraging age	Older	Younger
Body size	Larger	Smaller

*Modified from Winston (1987).

and her workers behind. Temperate swarms are more successful the earlier they form in summer – 'a swarm of bees in May is worth a load of hay . . .' (Opie and Opie, 1955), while tropical bees can swarm at any time.

Absconding is when all the adults in a colony take off, leaving the larvae behind. Such behaviour is generally non-adaptive for temperate bees. For tropical bees, faced with a wider range of enemies (ant eaters, armadillos, army ants, stingless bees, etc.) and occasional local failures of food, absconding can be adaptive, despite the loss of young. Similarly, stinging, which kills the worker, is less costly to tropical bee colonies with their more rapid population turn-over. So each set of characters can be seen to be adaptive in the right climate.

There are no native honey bees in the Americas. Brazilian bees were derived from temperate bees, German and Italian ones and their hybrids (Gonçalves *et al.*, 1991), and so were not well adapted to the local climate nor as productive as they might be. So the Brazilian Ministry of Agriculture decided that African bees should be imported. Dr W.E. Kerr, a bee geneticist, brought 170 *scutellata* queens from eastern and South Africa to Brazil in 1956. Some 47 or 48 survived importation, one from Tanzania, the rest from the Transvaal (about 27°S, so not strictly tropical). (Slightly different numbers are given by Winston, 1992.) Hives were set up at Campauâ, Rio Clara, São Paulo state. In early 1957 someone ('an ill-informed technician of the railroad' according to Spivak *et al.*, 1991; 'a well-meaning beekeeper' according to Gonçalves *et al.* (1991) and Winston (1992), could they be the same person?) took off the queen excluders and so allowed swarms to escape from 26 hives. Although part of the spread started with these escapes, some of the remaining African queens were bred and distributed, undoubtedly leading to other escapes.

The main biological argument about the spread is whether it is pure African or involves hybrids. Part of the difficulty is that, morphologically, it takes many measurements of many individuals, and a discriminant analysis, to distinguish African from European bees. In tropical lowlands, the selective advantage of the African behavioural and life history traits, and the Africanized bees' rate of increase, means that hybrids will be swamped even if they occur. Mitochondrial DNA showed that almost all the queens at the front of the advance were matrilineally African (Hall, 1991). Their behaviour also indicated that they were essentially African. However, there is undoubtedly hybridization as the front advances, as shown for instance by Rinderer *et al.* (1991) in Yucatan. Harrison and Hall (1993) have shown that the hybrids will often be at a selective disadvantage.

Where the terrain is less tropical, the picture is rather different. In Costa Rica (Spivak, 1991, 1992; Lobo, 1995) in lowland areas in the mid-1980s there was a rapid turnover from strongly European to strongly African. But at higher elevations there was a much slower establishment and the characteristics changed more slowly. Going south from Brazil, in Argentina there is now a transition zone running west from Buenos Aires and then north along the Andes. This zone is between the Africanized area to the north and the

European to the south, and is not much more than 100 km wide (Sheppard *et al.*, 1991). A comparable sorting out of subspecies by climate has been reported in Tasmania (Oldroyd *et al.*, 1995).

There is disagreement about what to call these bees, but Africanized seems to me the simplest and clearest, even if some tropical American populations may be more or less pure African genetically. They are not African geographically. The front has now reached the USA in Texas and Arizona (Rinderer, Oldroyd and Sheppard, 1993). If left to itself, a transition zone in the US like that in Argentina seems likely. But measures to move such a zone south are possible (Winston, 1992), if contentious.

The Brazilian experience shows that bee keepers can adapt to the African behaviour (Gonçalves *et al.*, 1991). The hives must be kept further apart and away from disturbance. With European bees, keepers can wear sandals, shorts and a simple face mask. For Africanized bees, keepers need light-coloured smooth boots, a light-coloured smooth overall with elastic cuffs with another layer of clothing beneath, a sturdy wire veil, pale on the outside and only black on the inside at the front panel, a light-coloured hat and gloves of light coloured plastic or rubber. They use a large smoker, 30 × 15 cm, with strong bellows, to control the bees. With these precautions, the number of hives in Brazil rose from 320 000 in 1972 to 1 980 000 in 1988. Honey production first declined from around 7000 metric tons per year in the 1960s to just over 4000 in 1974, but then rose steadily to 36 000 tons in 1988.

What does this example say of the role of genetics in invasion? Judging from the characters in Table 6.2 and the discussion above, there could be up to 100 genes involved, if all characters were purely polygenic. However, that has not been proved, and if major genes were found the number would be less. Either way, they act together as an adaptive complex, and certainly show the importance of having the right genes for a successful invasion.

6.3 GENETIC CHANGES DURING INVASION

6.3.1 Introduction

There are hints in the literature that successful invaders sometimes differ genetically from the stock from which they are derived. This may be by chance, or by unintentional selection if the species is maintained in captivity for some generations before being released. That is most likely with organisms intended for biological control, both because they have to go through a quarantine period to rid them of pathogens and parasitoids, and because it is usually desirable to breed up large numbers before release. There is also the possibility that there is rapid evolution from strong selection in the first generations after release (Hopper, Roush and Powell, 1993). This has been suggested particularly by biological control scientists, often using the term transilience.

It is difficult to assess these suggestions. As was discussed above, no genes have been identified, and breeding tests to demonstrate a genetic basis are lacking. Without such tests, in a 'common garden' for plants, or in similar controlled conditions for other organisms, it is normally impossible to know if a change in ecology should be ascribed to a change of habitat or a change of genes (Myers and Iyer, 1981). Transilience (Templeton, 1980) is a controversial theory (Barton and Charlesworth, 1984) devised by Carson (1975) to explain the evolution in the Hawaiian drosophilidae, particularly when that evolution was associated with the invasion of another island. In biological control, it has been used to describe unexpected population changes at first release (Murray, 1986).

Evolution over several generations after introduction is considered below (section 6.4) and might also be relevant to transilience. In this section the question is whether there is evidence for genetic change, transilience or otherwise, that enables the invasion to get going. The evidence is mostly that there is sometimes an eclipse period after the introduction of a biological control insect, a period in which the population disappears or is rare. A major success in biological control was the introduction of the chrysomelid beetles *Chrysolina hyperici* and *C. quadrigemina* to California in 1945 and 1946 respectively to attack (perforate) St John's Wort *Hypericum perforatum*. *C. quadrigemina*, the better control agent of the two, was not seen for three years (1939–1942) after its introduction to Australia (Wilson, 1965), and was scarce for 7–11 years after introduction to western Canada (Myers, 1987). *C. quadrigemina* is a Mediterranean species, taken from France to Australia, and from there to California, Oregon, Washington, British Columbia, Chile, South Africa and New Zealand (Wilson, 1965). There is no hard information about the causes of eclipse in Australia and British Columbia; it could be genetic, it could be an Allee (low density) effect, it could even be an ordinary slow population build-up (DeBach, in Wilson, 1965). The same doubt affects the other examples in Murray (1986) and Myers (1987).

All that is unsatisfactory and nebulous. The one area in which it is clear that genetic changes at invasion are important is hybridization. Hybridization can be important in invasions in two ways. The first is by producing a new genotype and so allowing an invasion of a new habitat. The invasion takes place as an (immediate) consequence of the hybridization. The second is that hybridization during invasion causes environmental problems. The first way is often shown by plants, and two examples are given here. The second implies some concern about genetic integrity, and, so far, is most common among vertebrates. Ducks, fish and mammals are the examples here, but this is a possible problem with all sorts of genetically engineered organisms, an aspect discussed in section 6.5 below.

6.3.2 Hybridization leading to invasion

The two examples of hybridization allowing the invasion of a new habitat are both grasses, *Poa* and *Spartina*. The *Poa* example is fairly trivial and is included to show that the well-known and important phenomenon in *Spartina* can be found elsewhere.

Poa alpina is a circumpolar species called alpine blue-grass in America, alpine meadow-grass in the British Isles, where it is a mountain plant, found in England, Scotland, Wales and Ireland. *P. flexuosa* wavy meadow-grass is much rarer, found in only four places (10-km squares) in Scotland (Perring and Walters, 1962), though it is more common in Norway (and just into Sweden) (Hultén, 1971) and also found in Iceland. *P.* × *jemtlandica* is the hybrid (the × signifies a hybrid), with the distribution of *flexuosa* (Perring, 1968), and apparently with a hybrid habitat. 'it may be significant that, in my experience at least, *P. flexuosa* chooses the open scree, *P. alpina* usually cliff-ledges, and *P.* ×*jemtlandica* the boulders or outcrops of rock on the edges, or even in the middle, of scree gullies' (Raven, in Raven and Walters, 1956). Curiously, *P. alpina* is nearly always proliferous in Britain, bearing plantlets instead of flowers or fruits, as is *P.* ×*jemtlandica* which otherwise looks more like *P. flexuosa*. Sexual forms of *P. alpina* have been recorded, but not in the area of the hybrids. All are perennial.

Poa ×*jemtlandica* is a minimal example of the ecological effect of hybridization allowing the invasion of a new habitat. *Spartina anglica*, another grass, is a much more spectacular example.

Spartina is a genus of 14 (Daehler and Strong, in press) perennial grass species which grow in salt marshes and on tidal mud flats. Most of the species are found on the North American Atlantic coast. There is only one European species, *S. maritima* small cord-grass, found from the Netherlands south to the western Mediterranean and Adriatic, and the Atlantic coast of North Africa. It has 60 chromosomes. In Britain it is confined to the south-east of England from the River Exe to the Wash with one record in south-east Ireland, and hardly ever seeds. Possibly it is introduced into north-west Europe (Gray, Marshall and Raybould, 1991). It occurs in south-west Africa which may be another introduction.

A common eastern American species is *S. alterniflora* smooth or salt-water cord-grass, with 62 chromosomes, found from Newfoundland to the Gulf of Mexico. It was introduced to Europe in the early 19th century. It was recorded from the Ardour estuary in south-west France in 1803, and from Southampton Water in the English Channel in 1829, though it may have been there as early as 1816 (Gray, Marshall and Raybould, 1991).

In Southampton Water the two species hybridized to give *S.* × *townsendii*, first described in 1879, a sterile species with 2n = 62. It was apparently first collected in 1870, though there is a description from 1861 that could refer to it (Gray, Marshall and Raybould, 1991). From isozyme evidence the origin is not in doubt (Gray, Benham and Raybould, 1990). In south-west France, *S. alterniflora* had to spread 40 km south to the River Bidassoa on the Spanish border before meeting *S. maritima*. *S.* × *townsendii* was formed there too, and described as *S.* × *neyrauti* in 1892. The hybrid was formed at the only two places in Europe that the parent species met. Both Marchant and Raybould did not manage to reproduce the cross in the laboratory, but the pollen of *S. maritima* is mostly infertile, at least in England. Hybridization in the wild is probably rare, as shown by the delay in hybrid populations being formed.

In Southampton Water a more vigorous plant was noted in 1892; extrapolation suggested it arose in 1890 (Goodman *et al.*, 1969). It was the allotetraploid, now called *S. anglica* common cord- grass, with 2n usually 122, sometimes 120 or 124 (Raybould *et al.*, 1991). Almost no isozyme variation has been found in *S. anglica*. What there is could be the result of chromosome loss. Again, it has not been formed again in the laboratory, so it is possible the whole species arose from a single plant (Gray, Marshall and Raybould, 1991). It is fertile and sets seed well. In recent years it has been hard hit by ergot *Claviceps purpurea*, a fungus that attacks the inflorescence. Probably this is a consequence of the genetic homogeneity and of the plant growing in large mono-specific stands (Plate 8). Ergot can attack up to 90% of the inflorescences and reduce fecundity drastically, but the population effects are not clear as seed germination is low anyway (Marks and Mullins, 1990; Gray, Marshall and Raybould, 1991).

Both *S.* × *townsendii* and *S. anglica* grow further down the shore than *S. maritima* and colonize bare inter-tidal mud (Goodman *et al.*, 1969). Mud builds up round the plants, and they may eventually be succeeded by the plants of the higher marsh. So they have been much planted to reclaim land, and much cursed for narrowing harbour channels. As the species were not distinguished for many decades, the history is a little confused. *S.* × *townsendii* may have only spread naturally along about 100 km of the south English coast (Perring, 1968), but it has been planted elsewhere in the British Isles and as far away as New Zealand. *S. anglica* is far more common as well as more vigorous. It grows in mono-cultures like a crop, and is found fairly continuously up to the south of Scotland and in many places in Ireland, well beyond the northern limit of *S. maritima*, and presumably would have spread naturally that far eventually. The most vigorous growth is between mean high water springs and mean high water neaps, but there are patches growing both higher and lower than that (Gray, Marshall and Raybould, 1991). It is now found in a few places in the north of Scotland (Goodman *et al.*, 1969). As it has often been planted it is not clear what the natural range would be, but it seems likely that it would extend appreciably further north than *S. maritima*.

On the east coast of North America, *S. alterniflora* also grows in the low marsh. It would grow higher but for competition from *S. patens* (2n = 28, 35, 42), variously known as salt-meadow grass, high-water grass or musotte. *S. patens* cannot grow in the low marsh because of the high salinity and anoxic soils (Bertness, 1991). The upper limit of *S. alterniflora* is about mean high water, lower than *S. anglica* but that is probably the result of competition with *S. patens* in one case but not the other. The lower limits are not very different.

On the west coast of America there is a native species *S. foliosa*, probably an allospecies of *S. alterniflora* but with a narrower vertical range. *S. alterniflora, anglica* and *patens* have all been introduced, as has a South

American species *S. densiflora* (Daehler and Strong, in press). It is too early to be certain what will happen, but *alterniflora* may well displace *foliosa* in California and be displaced by *anglica* in Washington State. In San Francisco Bay, *alterniflora* is doing well but *anglica* occupies less than a hectare. *S. patens* will presumably divide the habitat as it does on the east coast, but the effect of *densiflora* has not yet been predicted.

The ecological effect of forming the hybrid *S. anglica* seems to be greater than just a hybridization of habitat. It combines the habitat preference of its American parent, but possibly goes further down shore, with the tolerance of cool summers of its European parent, but can go considerably further north. In this case hybridization has allowed a remarkable new invasion.

Spartina anglica is a notable allotetraploid invader, with an ecological amplitude different from both its parents. That last point is true of many other allotetraploids. Another grass example is cocksfoot *Dactylis glomerata* (2n = 28), a common meadow grass of north-west Europe. Its parents are *D. polygama* (= *aschersoniana*, 2n = 14) a woodland grass of central Europe, and *D. woronowii* (2n = 14), a steppe species from further east. The tetraploid is found not only in a new habitat but also, in some parts of its range, in the ancestral habitats (Stebbins, 1971). No doubt cocksfoot was once an invader too.

6.3.3 Hybridization at invasion

The other important genetic effect that happens at invasion is hybridization leading to genetic fusion. This is known in several species of birds, fish and mammals. The main example I will use is of the stiff-tailed ducks, *Oxyura* (Figure 6.3).

One of the remarkable features of species distributions is that they have quite well-defined boundaries, and that there are areas beyond those boundaries where the species does not occur despite what looks like usable habitat. The world map of *Oxyura* shows this (Figure 6.4). They occur on every continent, but patchily. As a genus, they occupy a wide variety of climates, yet there are many places, such as north-west Europe, where they are not found

Figure 6.3 White-headed duck, *Oxyura leucocephala* and ruddy duck, *O. jamaicensis*. Lengths about 45 cm. These are males, drakes. (From Harrison, 1982, with permission from HarperCollins Publishers.)

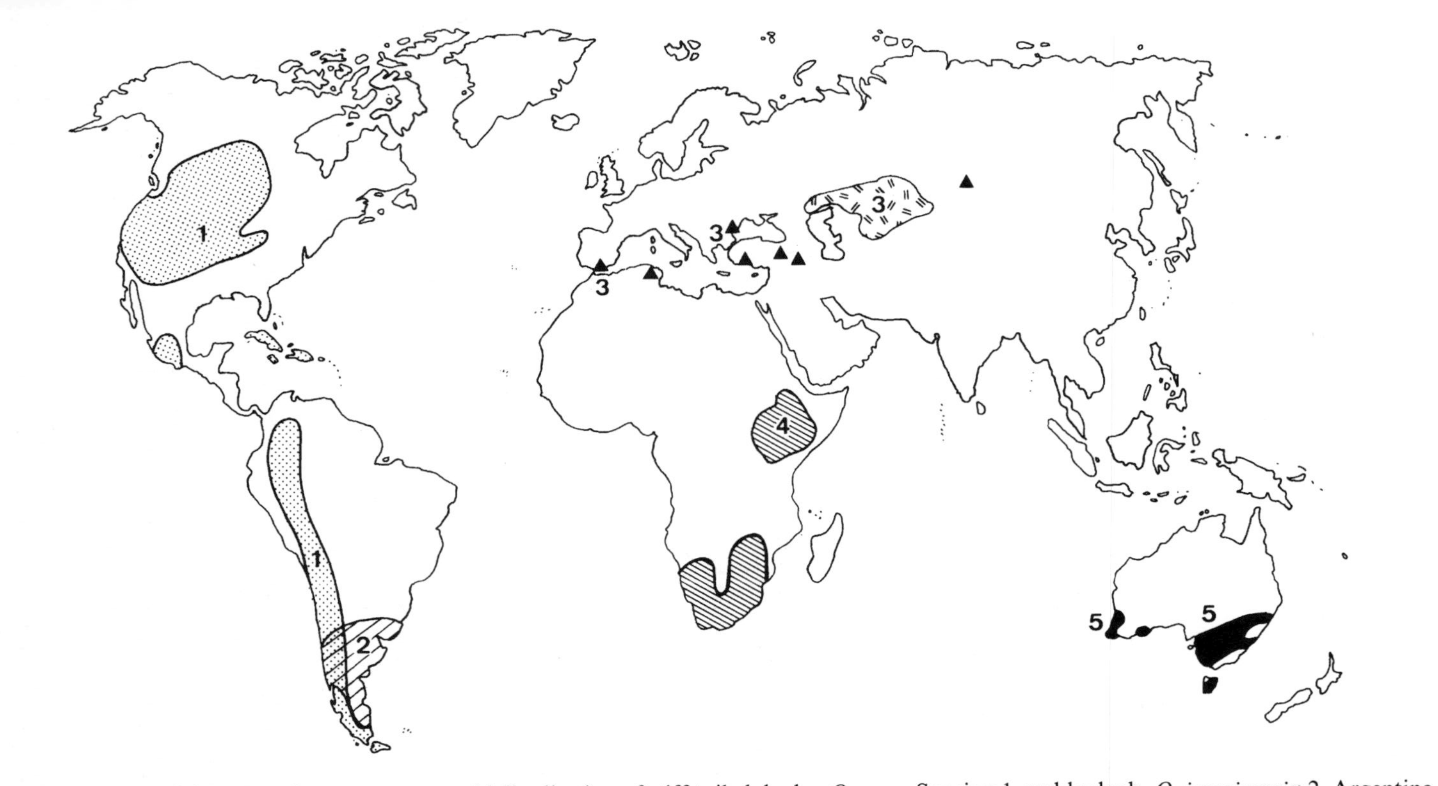

Figure 6.4 Map of the approximate present world distribution of stiff-tailed ducks, *Oxyura*. Species: 1, ruddy duck, *O. jamaicensis*; 2, Argentine blue-bill, *O. vittata*; 3, white-headed duck, *O. leucocephala*; 4, Maccoa duck, *O. maccoa*; 5, blue-billed duck, *O. australis*. (Illustration by Mike Hill.)

native. There is no close ecological equivalent in northern Europe. *Oxyura*, which means sharp-tail, though they are usually called stiff-tails, are dumpy diving ducks living in shallow lakes. They are poor walkers and lay remarkably large eggs.

One species of *Oxyura*, the masked duck, *O. dominica*, is rather different from the rest and overlaps geographically with two of them, the ruddy duck, *O. jamaicensis* and the Argentine blue-bill or lake duck, *O. vittata*. Those latter two overlap slightly in South America without interbreeding, but otherwise they and the other three taxa are almost allopatric. Were it not for that overlap, all five could conceivably be called subspecies. The distinction between a species and a subspecies is, in birds, the breeding one. Distinct species will not breed together, subspecies will. In practice, the distinction is often less absolute, and judgement is called for. The males, the drakes, of these five species are reasonably distinct, but the females, the ducks, are very similar.

There is concern about hybridization between ruddy ducks, *O. jamaicensis*, derived from North America and the white-headed duck *O. leucocephala*. In North America the ruddy duck is primarily a breeder in western central parts, though there are isolated populations elsewhere (Godfrey, 1966; North American breeding bird survey, internet http://www.im.nbs.gov/bbs/bbs.html). Three pairs were imported to Sir Peter Scott's Wildfowl Trust (now the Wildfowl and Wetlands Trust), Slimbridge, in 1948. Ducklings were difficult to rear in captivity; better results came from allowing the parents to rear their own young. Consequently some young avoided pinioning and escaped. The first feral breeding was reported in 1960 (Sharrock, 1976). Since then the British population has grown fairly steadily, reaching about 3500 in January 1992 (Gibbons, Reid and Chapman, 1993), increasing at about 15% each year. All this was thought a great success, an addition to British biodiversity. Sir Peter Scott managed to get the name of a pub near Slimbridge changed from the Black Bull to the Ruddy Duck (Kear, 1990).

The white-headed duck has shown a relic distribution for some time (Voous, 1960) and has become endangered. In this century breeding populations have been lost in Albania, Egypt, France, Greece, Hungary, Italy, Morocco and Yugoslavia, largely through loss of wetlands and shooting (Marti, 1993). The Spanish population was down to only 22 birds in 1977, but strong conservation measures brought this up to about 800 in 1993, a wonderful success story for Spanish conservationists. There are still breeding populations in Algeria and Tunisia. The main population of the species, perhaps 18 000 birds, is to the east, in Kazakstan (Figure 6.4).

Meanwhile, the ruddy duck has spread from Britain to Europe, and has bred in Iceland, the Netherlands, France and Spain. It was first seen in Spain in 1983, and in Andalusia in the range of the white-headed duck in 1986, though the first breeding record was in 1991. The ruddy duck is the more aggressive, competing for resources and mates. By the end of 1993, 16 ruddy ducks and 34 hybrids had been shot by marksmen working with conservationists (Holmes, 1994). The Spanish are rightly alarmed at the prospect of an endangered species being eliminated both ecologically and genetically by an invader, and probably would not be much less alarmed if the invader were called a subspecies.

In response, a UK Ruddy Duck Working Group was established in 1992 to see what action could be taken. The European Union will try to act under directive 79/409/EEC. Whether effective action to limit the British and mainland European ruddy duck population is possible is not yet clear (Holmes, 1994). Corbett (1994) says such action is neither desirable, feasible nor ethical, as it involves killing one species to preserve another which is not actually being killed, only (probably) out-competed. He suggests that, if the intention were to preserve stiff-tails, then the hybrid might be the taxon of choice. Hawkes (1993) calls killing birds of one species to save the genes of a closely related one 'an odd bargain'. Nevertheless, probably most conservationists would regard the preservation of local races, of local genotypes, to be paramount.

There is no doubt that many naturalists believe strongly that native genes should be preserved, and that introducing foreign genes is genetic pollution. The same issue arises in sowing wildflower seed of Continental (and sometimes agricultural) origin to recreate British meadows (Akeroyd, 1994). In that case, it has long been thought good practice to use seeds from the most comparable population; it is commercial availability that leads to other seeds being used. With stiff-tails there are clearly good arguments both for preserving the original Spanish genotype and for allowing the unrestricted spread of a species which is, in most of Europe, creating new diversity; regrettably there is no compromise solution.

Other hybridization problems raise the same issues. Still with ducks, the mallard *Anas platyrhynchos* is both one of the world's most common wildfowl species and the species best adapted to a range of anthropogenic habitats. It is the ancestor of the ordinary farm duck. But there are round the world a number of taxa, usually given specific rank, with which it can hybridize. In Australia and New Zealand, where the mallard is introduced, it hybridizes with *A. superciliosa* (Pacific black duck in Australia, grey duck in New Zealand), but particularly near towns, much less in rural areas. The increase of the mallard is fairly slow in Australia, and probably helped by new releases (Blakers *et al.*, 1984), but causes concern. In New Zealand, although introduced in 1867, it did not become established until there was intensive breeding and liberation in the 1930s. It now outnumbers grey ducks in closely settled districts and highly developed farmlands (Falla *et al.*, 1978). However, Gillespie (1985), from careful morphometrics, suggested that there were both grey-like and mallard-like hybrids, and that fairly soon there would just be a complex hybrid swarm, even though individuals might appear to belong to one or other species.

In North America, close relatives of mallard are Mexican duck *diezi*, mottled duck *fulvigula* and black duck *rubripes*. Mayr and Short (1970) list the first two as subspecies of mallard, though many American bird books call them species. The black duck is usually, but not always, ranked as a species (Voous, 1973), and it and the mottled duck are unusual in North American ducks in having a wholly eastern distribution. 'The Mexican duck has been removed from the US endangered species list after the discovery that all 5000 of the US population are hybrids between Mexico's 50 000 pure Mexican ducks and the common mallard' (Anon., 1978). In Canada, mallard were common breeders in the western provinces, progressively scarcer further east (Godfrey, 1966). However, mallard are increasing in the east and appear to be causing a decline in black duck; both hybridization and competition may well be involved (Ankney *et al.*, 1987). Hybrids certainly occur quite commonly.

Similar problems are known with fish. For salmonids, salmon and trout, Hindar, Ryman and Utter (1991) show that at least nine species have been affected by releases or escapes of cultured stocks. The outcome varied from no detectable effect to complete introgression or displacement, but, more ominously, the effects always appeared to be negative. 'We recommend strong restrictions on gene flow from cultured to wild populations and effective monitoring of such gene flow'. *Gambusia affinis*, a poecelid fish, is often introduced as a biological control for mosquitoes. It is yet another case showing that non-specific biological control agents should be banned. It is a pest (Welcomme, 1992), harms native species (US Congress OTA, 1993) and has hybridized with a restricted endemic *G. heterochir* (Courtney and Meffe, 1989).

Hybridization is also known in mammals (Lever, 1994). For instance, sika deer *Cervus nippon* have created a hybrid zone with red deer *C. elaphus* in Scotland, a zone which is moving and not at equilibrium (Abernethy, 1994).

6.4 GENETIC AND EVOLUTIONARY CHANGES AFTER INVASION

6.4.1 Genostasis

Hybridization can lead to rapid genetic change at invasion. Do other genetic changes also happen rapidly? The answer seems to be, on the whole, no. Invasions are fast, evolution is slow. First, I will discuss the evidence that genetic changes are often not detectable in ecological time, that is, in less than a century. Then come some cases in which genetically minor, but ecologically important, changes have been detected. Lastly, a brief look at some evolutionary changes over longer periods, and the implications those have for expectations of genetic change after invasion.

The distribution of *Oxyura* ducks (Figure 6.4) drew attention to the common phenomenon that species seem often unable to spread beyond their boundaries into apparently suitable habitat. To put that another way, there are limits to the climatic adaptation of species. Biological control has led to many natural experiments that test this. Wilson (1965) pointed out that on the whole evolution did not happen after introduction, at least not detectably in a short period. 'Generally, the natural enemies used for biological control seem to be characterized by fixity of behaviour rather than adaptability. They usually operate effectively only in a part of the area occupied by the host. . . . Biological control experience seems to indicate that adaptive evolution in introduced insects generally occurs at a rate that must be very slow judged by the time scale of biological control research . . .'. He goes on to instance a few cases of apparent evolution, and I return to some of these below.

Mineral soils from mining may produce many patches of a new sort of habitat. Invasion of these patches often requires genetic adaptation, for example to high concentrations of copper. Bradshaw (1991) noted how few species seem able to adapt. Putting this evidence with Wilson's and other cases, he emphasized the phenomenon which he called genostasis. 'Most species are very stable . . . Evolutionary failure is commonplace'. In Bradshaw's view, the primary reason for genostasis is the lack of usable genetic variation. Nowadays it is well known that most species contain much genetic variability. Sometimes this allows rapid micro-evolution, but Bradshaw's point is that these are the exceptions. Mostly, whether in natural situations, or in the development of plants for agriculture and horticulture, despite the variation, the particular genes wanted for an adaptive trait are missing. He points out that this is one reason why companies are interested in genetic engineering: it allows them to overcome genostasis.

Another indication of the importance of genetic variability in limiting invasions is an experiment on rose clover *Trifolium hirtum* (Martins and Jain, 1979). Attempts to found colonies outside its invasive range in California were more successful with high levels of genetic variation.

Now that the detailed molecular make-up of genomes is becoming known, it should soon be possible to examine molecular reasons for genostasis. With the genes present in an organism only a finite, if fairly large, number of functional mutations is possible. Forming a new gene for a particular function may be impossible. Perhaps this is one reason for the success of hybrids. The two species involved have different adaptational abilities. Putting their resources together gives access to new adaptations.

6.4.2 Recent genetic changes

One change that can follow invasion would certainly reinforce genostasis. That is the loss of genetic variation. An example is *Andricus quercuscalicis*, a small (2-mm) gall wasp (Hymenoptera, Cynipidae). Like many cynipids it has an alternation of generations, with an asexual generation on the acorns of pedunculate oak *Quercus robur* and a sexual generation on the male flowers of Turkey oak *Q. cerris*. Turkey oaks have been introduced into western Europe far beyond their native range of Hungary, Italy, the Balkans and Turkey. Pedunculate oak is found throughout the British Isles and western Europe and north into Scandinavia (Hultén and Fries, 1986), native since the ice age in most if not all of this range.

Historical records allowed Stone and Sunnucks (1993) to map the spread of *A. quercuscalicis* from central Europe in the 17th century, to most of Germany by the end of the 19th century. They have a record for western France from 1940, and the species was first found in Britain in 1958. By 1991 it had spread to the east of Ireland and the south of Scotland. The spread has been marked by a decline in allele diversity and mean heterozygosity. Differences in allele frequency between populations were substantially greater in the invaded than in the native range. The evidence is that the loss came from the small size of invading populations rather than from selection. They map the loss of various alleles across continental Europe and Britain. There were no gains, and the losses were progressive and cumulative.

Other insect populations that have shown similar losses are listed by Stone and Sunnucks (1993). They include cockroaches and various Diptera. In birds, bottle necks, founder effects and other causes of loss were found in mynahs *Acridotheres tristis* (Baker and Moeed, 1987), starlings *Sturnus vulgaris* (Ross, 1983), house sparrows *Passer domesticus* (Parkin and Cole, 1985), and, though not statistically significant, tree sparrows *Passer montanus* (St Louis and Barlow, 1988). Such loss is also known in *Sarracenia* (pitcher plants) and *Theba* (land snails) (Stone and Sunnucks, 1993) and would seem to be the most common genetic effect associated with invasions.

Nevertheless, there are a sprinkling of cases in which genetic innovation has, it seems, followed invasion. Great mullein, *Verbascum thapsus* (section 3.4.6) may be an instance. The evidence is mostly indirect. Some cases involve adaptation to a new food source, and in these it is particularly difficult to distinguish genetic from purely phenotypic change, but it can be done. Most of Wilson's cases are of this type. An example of the sort of evidence, and its limitations, is Goeden *et al.*'s (1986) study of weevils on thistles in California. Both the weevils and the thistles are invaders from Europe. *Rhinocyllus conicus* (Coleoptera, Curculionidae) has various biotypes associated with different thistles. The main study was on the *Carduus pycnocephalus* Italian thistle (or Plymouth thistle) and *Silybum marianum* milk thistle biotypes. The larvae of the Italian thistle biotype were unable to develop, or apparently to evolve to develop, on milk thistles. The milk thistle biotype increasingly over a decade also attacked Italian thistles. The increase was from 0.1% attack in 1973 to 47.8% in 1983. The only genes available for study were a few identified by electrophoresis. While these showed that the biotypes had different frequencies of many alleles, a few alleles unique to one or the other were also found. On that evidence, the biotypes are probably sibling species. While the genes found are probably irrelevant to the widening of the host range, the phenomenon is nevertheless probably a genetic one.

Widening of the host range by genetic change after a species has been introduced for biological control is a topic deferred from section 5.2.1. *Rhinocyllus*, while probably an example of that, is more clearly an example of a common phenomenon in biological control, the predictable attack on non-target organisms. It also feeds on several native *Cirsium* species (also thistles) in California (Turner *et al.*, 1987). Simberloff and Stilling (in press) give other examples, which reiterate the point that biological programmes have often ignored non-economic species.

Even when a screening programme shows that a biological control agent is specific for a pest species, evolution can undo that specificity. Dennill *et al.* (1993) found that an Australian pteromalid gall wasp, *Trichilogaster acaciaelongifoliae*, imported into South Africa to control the Australian long-leaved wattle *Acacia longifolia*, was attacking two other introduced Australian acacias. These were blackwood *A. melanoxylon* and stink bean *Paraserianthes lophantha*, which are both weeds in parts of South Africa. Blackwood is also cultivated for fine furniture wood. The wasp was not found to attack blackwood in Australia, and did not attack it in host-specificity trials in South Africa. So there has probably been a genetic change. The damage to blackwood is actually minor.

Dennill *et al.* (1993) said that about 20 similar shifts were known in weed biocontrol programmes, but do not list them. As specificity is the characteristic that makes biological control an acceptable strategy, so loss of specificity casts doubt on the wisdom of that strategy except when there is no similar species to become a target. Given the generally slow rate of evolution and the permanence of biological control, these host shift cases indicate that further thought is needed on how to regulate biocontrol programmes.

Pesticide resistance is probably the most common class of reported genetic adaptations, and will occur in invasive species as much as native ones. One example is *Aphytis melinus* (Hymenoptera, Aphelinidae) imported from Pakistan and India (Simberloff, 1986) to control red scale *Aonidiella aurantii* (Homoptera) in California, which has developed insecticide resistance both naturally and under artificial selection (Roderick, 1992).

Climatic selection and adaptation is perhaps more interesting. That is what is probably involved in the minor morphological changes in house sparrows in America (Johnston and Klitz, 1977). The genes involved are not known. Some possible cases in biological control insects are given by DeBach (1965), the evidence being from range change. The eclipse of *Chrysolina quadrigemina*, the St John's wort beetle, as it spread north into British Columbia was noted above (p. 161). This beetle varies in colour, showing a range of purples, blues and greens as well as a distinct bronze morph. Peschken (1972) found the Canadian populations had a lower proportion of bronze, laid more eggs and sought shelter in the bases of the plants more than those in California. However, *C. quadrigemina* has not been able to adapt to the harsher winters in the interior of British Columbia, though *C. hyperici* can. Unfortunately, *C. hyperici* by itself fails to control the weed (Williams, 1986; Myers, 1987). *C. quadrigemina* seems to have shown a limited climatic adaptation (Murray, 1986).

While genostasis may be the rule, fairly minor genetic but important ecological changes are possible in some cases. This is consistent with much of the long-term and geological record, discussed next.

6.4.3 Longer-term changes and speciation

One possible eventual consequence of invasion is speciation, but as yet our understanding of the number of genes involved and the rate of evolution are fragmentary. The rate of evolution should be perhaps measured by gene substitutions (or additions) per unit time. It is well known that taxonomic, morphological and genetic evolution can proceed at different speeds. Appropriate measures of the speed of evolution have been discussed at least since Simpson (1944).

Evolution from invasions is well illustrated by double invasions of islands (Williamson, 1981). An example is the two species of chaffinches *Fringilla* on the Canary islands, in the Atlantic off north-east Africa. The older species is the blue chaffinch, *F. teydea*, confined to pine forests on the two largest islands. As Hooker, in 1866, pointed out is usual in such cases (Williamson, 1984), it is the rarer of the two. The other is the common European, *Fringilla coelebs*. There is insufficient information on the timing of the two invasions, or the genetic differences, to say anything useful about the rate of evolution, though evolution there must have been. Both are morphologically distinct from mainland populations.

A point often overlooked, and related to genostasis, is that many forms fail to evolve at all. Coope (1990) reiterates the point he has often made, that no morphological evolution can be seen in European Pleistocene beetles. That is an assemblage of thousands of species over more than a million years (Elias, 1994).

Interchanges between continents or between oceans offer many examples of invasions over a geological time scale, and the evolution that followed. Vermeij (1991a) summarizes a dozen and (1991b) discusses one in detail, the interchange of shell-bearing molluscs between the Atlantic and the Pacific via the Arctic ocean in the last 3.5 million years. Of 261 species found in the Atlantic and originating from the Pacific, 99 are derived, have evolved to new species, 162 are not, are still conspecific with the Pacific populations. The other way, in the Pacific originating from the Atlantic, 10 are derived, 24 are not. Most species do not evolve. About three in eight speciate in a new environment in 3.5 million years.

Those that evolve would seem to take, on average, something of the order of a million years to speciate, that is, less than 10 million years but more than 100 000 years. The upper limit is given by the geological period, the lower by noting the number involved. This estimate seems to match Coope's beetles too, and is consistent with the low rate of evolution seen in modern invasions. Knowlton *et al.* (1993) found seven lineages of snapping shrimps *Alpheus*, separated into Pacific and Caribbean populations by the rise of the Panama isthmus, had all speciated in 3 million years or more. Again, this seems consistent with the view that speciation commonly has a typical time scale of a million years.

Vermeij (1991a) emphasizes that interchanges are often asymmetric. That can be seen in the number of species in his Atlantic/Pacific example, though the speciation rates are remarkably similar (and certainly not statistically different, on a chi-square test). In the Great American Interchange of mammals between North and South America (Webb, 1991) there is an asymmetry in speciation. Invaders of South America speciated much more than those going north. Even so, the timing and the numbers suggests to me that speciation in less than 100 000 years is unusual. It is not surprising that we do not see it with modern invaders, and we still would not even at two orders of magnitude faster.

All that is imprecise and insecure as far as rates are concerned. A better measure of rapid evolution rates comes from another island example. On many islands, large mammals, such as elephants and deer, evolved to smaller sizes, small mammals to larger ones (Lomolino, 1985). Some extant examples survive, such as the 'toy' whitetail deer, *Odocoileus virginianus clavium* on the Florida Keys (Burt and Grossenhieder, 1976; Hall, 1981). Davis (1987) reviews the record of Mediterranean islands, Roth (1992) of elephants world-wide. Two examples allow some estimate of the rate of rapid evolution in large mammals. These are mammoths on Wrangel island in the Arctic ocean, near the Bering strait, and deer on Jersey, in the Channel Islands off the north coast of France.

The Jersey data are the more precise. Lister (1989) estimated that red deer *Cervus elaphus* shrank to one-sixth of their weight in less than 6000 years during the Eemian interglacial, around 120 000 years ago. That is a linear reduction of about 45%. Vartanyan, Garutt and Sher (1993) found teeth of both normal-sized and dwarfed mammoths *Mammuthus primigenius*. The linear reduction was 30% or more, the time again about 6000 years, from 12 000 years ago to between 7000 and 4000 years ago. The Jersey evolution is apparently the faster, but is still only a linear reduction of about 1% per century. If 1% per century is, from the geological record, fast, it is inevitable that evolution in characters determined by several genes will not normally be detected in the first few years or decades after an invasion.

Invasion is fast, evolution is slow. What all these examples indicate is that evolution will take place after invasion, but often at a speed too slow to be measured. In some cases, significant ecological change will happen in tens of generations, but these will be the exceptions. An important brake on the rate of evolution may well be the availability of new genes, of genostasis.

6.5 THE RELEASE OF GENETICALLY ENGINEERED ORGANISMS

There is some dispute about whether the study of invasions is relevant to the assessment of risks from the release of genetically engineered organisms. So, first, what are these organisms and why should there be risks?

Molecular genetics now allows a gene to be taken from one organism and inserted into some totally unrelated one. Bacterial genes can be put into plants, arthropod genes into viruses. There are, of course, limits on what can be done, but, as the subject is moving fast, I will not dwell on them here. This transfer of genes is commonly called genetic engineering. The current fashion is to refer to genetically modified organisms, or GMOs, rather than genetically engineered ones, and that is followed in official documents. For scientists, there is no reason to prefer an ambiguous and obscure term to one that is reasonably precise (Williamson, 1992). Genetic modification is a term that can be applied to all genetic programmes, and has no obvious association with restriction enzymes and the other tools of molecular geneticists. Genetic engineering is less ambiguous, and gives the flavour of experimental manipulation, so I will use it here. It is also the term used in US Congress OTA (1993).

Genetic engineering can be used to make organisms with new properties that may be commercially useful. In crop plants, herbicide- and pest-resistant varieties are being developed in many species. It is possible to change the nature of the crop product, changing the composition of the oil in oil seeds, manipulating enzymes so that tomatoes do not go squashy, and many other features. Pharmaceuticals could be made in plants or produced in milk. Fish can be made to grow faster (US Congress OTA, 1993; Krattiger and Rosemarin,1994).

In principle, any commercially desirable trait could be added or enhanced. It is not surprising that much research has been funded, and that many commercial releases are near. On Krattiger's (1994) count there have been 2053 field trials of transgenic plants world-wide up to mid-1994, and that, even allowing for differences in the definition of a trial, is fairly certainly too low (Anon., 1994b). Although almost all OECD countries have some form of regulation, others outside the OECD such as China and Israel apparently do not. Regulation, such as the European Union's directive 90/220/EEC, is likely to keep only a few, rather obviously undesirable, products from the market. It is reasonable to assume that there will soon be many different genetically engineered organisms marketed in large numbers world-wide.

Will there be ecological and environmental change from genetically engineered organisms? Russo and Cove (1995) give a good overview of all the benefits and hazards from these techniques. Invasions show that damage can happen when an organism finds itself in a new environment. For a novel genetically engineered organism all environments are new. A familiar case where a change to a new environment, accompanied by a small genetic change, has had quite unforeseen terrible effects is AIDS. Maybe someday a genetically engineered organism will produce a major, but quite different, disaster.

Human AIDS is caused by two viruses, HIV1 and HIV2. These are closely related to a group of viruses found in other primates, the Simian Immunodeficiency Viruses or SIVs (Morrison and Desrosiers, 1994). These are all single-stranded, encapsulated RNA viruses, retroviruses. Being RNA viruses, they are far less stable genetically than DNA organisms. Various strains in one virus have 80–100% identity, closely related retroviruses attacking other species have about 80–90% identity, more widely related ones 55–60%. It would seem that both HIV2 and SIV_{mac} (which infects captive macaques, *Macaca*) are derived from SIV_{smm} (which infects the sooty mangaby, *Cercocebus torquatus*). Similarly, HIV1 is closely related to SIV_{cpz} which is found in chimpanzee *Pan troglodytes*. SIVs in wild monkeys and apes are, as far as is known, non-pathogenic. Rhesus monkeys *Macaca mulatta* with SIV_{mac} develop an AIDS-like disease. Pigtail macaque *M. nemestrina* with the same virus are killed in a week or so. Small genetic changes and a new environment can produce very drastic effects.

As I said at the end of section 5.3.1, major invasions may come out of the blue at any time. Will genetically engineered organisms add to these problems?

Some proposals, such as the engineering of non-specific biological control viruses, are evidently bad practice (Williamson, 1991), but the unnecessary risk comes from the nature of the virus, not the genetic engineering. It is often asserted that for most commercial genetic engineering, the invasion model is not appropriate. For instance, with crop plants, the argument is that the plant is familiar, the new variety will have to undergo extensive performance trials, and the genetic novelty is more precise and better understood than the genetic novelty produced by traditional breeding programmes. Hence the release of genetically engineered plants is different from other invasions. It is also sometimes stated that many changes are needed to change an organism into a weed or a pathogen (National Academy of Sciences, 1987).

The unsatisfactory points in that argument are covered in earlier sections of this book. Although the crop plant is familiar, and the genes inserted are well known, the combination is novel. There are no universal characters that distinguish weeds and pathogens from their harmless relatives (section 3.3.2), and the genetic differences between invasive species and those that fail to invade may often be small (section 6.2). In fact, many plants have become weeds merely by being taken to new regions. It is not surprising that ecologists think that aspects of the invasion model are relevant to the risk assessment of genetically engineered organisms (Tiedje *et al.*, 1989; Altmann, 1993; Shorrocks 1993; US Congress OTA, 1993; Seidler and Levin, 1994). It is an appropriate model.

Even in those countries where there is effective regulation of small-scale trials, the study of invasions suggests that the probability of detecting undesirable products at an early stage is not large. Many pest invaders have not been recognized as such for many years, often decades, *Impatiens glandulifera* (section 1.3.3) and the muntjac deer (section 5.1) for example. On the other hand others, such as zebra mussel (section 5.4) were recognized as problems almost immediately, but spread so fast that it was difficult to limit the damage. As those genetically engineered organisms that become problems will usually be commercial products, they will mostly be widespread quickly, and difficult to control whether the problem arises soon or not.

Some problems may well be delayed. Texas cytoplasm is a possible example of how this could happen; it is a genetic modification of corn, *Zea mays*. In corn, male sterility is a most useful agronomic trait, because it allows controlled breeding without the work of removing the male inflorescences, the tassles. Texas cytoplasm varieties are male sterile because of a change in a mitochondrial gene (Levings, 1990). Remarkably, the same molecular mechanism that made the plants male-sterile also made the plants susceptible to a fungal pathogen, *Bipolaris maydis* race T. About two decades after the gene, T-*urf*13, came into commercial use, the fungal disease devastated the corn containing that gene, which was by that time 85% of the US corn hectarage, and made the innovation useless. The molecular details were known, the pathogen was known, but the interaction was not predicted and the consequences did not appear until the new genotype was in full commercial use. It seems optimistic to suppose that similar failures will not happen in future, however the regulatory system is designed. Without regulation they might even become common.

Texas cytoplasm was an agronomic problem. Will there be ecological and conservation problems? There are two classes of possibility. One is the spread of the genetically engineered organism itself, the other is the spread of the engineered gene in wild relatives of that organism (Raybould and Gray, 1993). As with all invasions, and bearing the tens rule in mind (section 2.3), it is reasonable to say that neither will happen frequently. Pests arise in around 1% of organisms introduced at random. If regulators can control excessive commercial enthusiasm, the frequency could be much less (Williamson, 1988); that is taking an optimistic view of the effectiveness of regulators and regulations. But whatever the proportion, the number of proposed products is so large, that some ecological damage seems likely, though it may not be apparent for some decades.

Perhaps the most remarkable general feature of invasions is how unpredictable they are. One possible gain from the release of genetically engineered organisms may be a better understanding of why most genetic variation seems to have no relation to invasion success, but nevertheless some genes are important.

7 Implications and communities

Conceptual framework points (from Table 1.1):

CFP 9 Invasions are informative about the structure of communities and the strength of interactions, and vice versa

CFP 10 Invasion studies are relevant to considering the risks of introducing new species or genotypes, the release of genetically engineered organisms, and the success and consequences of biological control

7.1 INTRODUCTION

Up to now, the emphasis has been on individual species, their probability of success, their rate of spread and their genetics. This chapter is about the effects of invasions on communities, about how the ecological community as a whole is changed by invasion. With that, I will have visited what I consider the main problems in the study of invasions, and it will be time to give a brief overview of the main conclusions.

No man is an island (Donne, 1624), and every species interacts with some others. All animals require something to feed on, all plants have species that feed on or in them. As well as these universal interactions, there are common, interesting but less universal ones, like symbiosis or habitat alterations that affect other species. Any successful invasion must have some consequences for the other species present. But the main lesson from Chapters 2 and 3 is that most such effects are minor. That in itself tells us something about the working of ecological communities.

Some major consequences of invasion were discussed in Chapter 5. A distinction was made there between ecosystem effects (section 5.5) and others. But all such major consequences have measurable effects on other species. The most severe effect is the extinction of other species. Some invasions have resulted in the extinction of many other species. Examples are *Boiga* (introduced accidentally) or *Euglandina* (introduced for biological control) (section 5.6). In Chapter 5 the focus was on the invading species;

here the focus will be on the effects on the ecological community that is invaded.

There are three relevant sets of studies. The first is a set of accounts of invasions that are notable for the way they affect many species, invasions with so-called ecosystem effects (section 7.2.). Second, some mathematical, computer-based, studies of model ecosystems (section 7.3). The third set (section 7.4) is examples that show the effect of introducing many species, or cascades of effects after an introduction, or both. This third set in particular tells us both about the effect of invasions on communities, and also indicates something about the structure of communities. All that leads to the still contentious issue of which communities are more susceptible to invasions than others and why.

7.2 ECOSYSTEM EFFECTS

There are many different ways in which an invader can affect many other species. One of the joys of ecology and biology is the variety and unexpectedness in the ways that organisms make their living, but a consequence is that ecological effects can seldom be catalogued neatly in distinct classes. Vitousek (1990) listed and classified some effects of invaders, Macdonald *et al.* (1989) gave an overlapping and rather wider list. Some of the Macdonald *et al.* examples fit a little uneasily in Vitousek's categories. Here I will try to combine the two lists but use the Vitousek categories, as they are a logical set. Several of the examples were discussed in Chapter 5, but need to be mentioned again here to give context to the following sections.

The three Vitousek categories were: (1) effects from resource acquisition or utilization; (2) effects from altered disturbance frequency or intensity; and (3) effects from altered trophic structure. There was a fourth, null, category: 'I believe that the majority of successful invasions do not alter large-scale ecosystem properties and processes in a meaningful way' (Vitousek, 1990). The meanings are most simply explained by examples.

Resource effects can include effects on the flow rates and directions of nitrogen, water or other inorganic chemicals, or on more complex, but still not biological, resources. These were part of the ecosystem effects in section 5.4. Examples included *Tamarix* spp. and water, *Myrica faya* and nitrogen, and *Mesembryanthenum* and salt. Another effect of *Mesembryanthenum*, because it makes the soil surface salty, is to suppress other plants and increase soil erosion. Other plants such as *Ammophila arenaria* marram grass on sand dunes or *Spartina anglica* (section 6.3.2) on tidal mud flats fix soil (in a wide sense), rather than increase erosion. For that matter, *Salvinia* (section 5.3), by making a dense mat on otherwise open water, could be said to be sequestering a resource.

Soil erosion shows the difficulty in distinguishing categories, as it involves both a depletion of a resource and an increase in disturbance. The examples of disturbance that Vitousek (1990) used were, first, invasive grasses that ıcrease the frequency of fire in semi-arid shrublands, such as cheat grass *Bromus tectorum* in western North America, and, second, rooting by feral pigs, or instance in Hawaii, as described in section 5.5. Both fire and rooting are auses of concern in several places round the world.

On the boundary between disturbance and a trophic effect is the suppression of recruitment of native plants by invasive ones. Such swamping was iscussed in section 5.4 mentioning several plant examples, the water fern *Salvinia* and zebra mussel *Dreissenia*. Some other examples are shrubs and ines. Rhododendron *Rhododendron ponticum* in the oak woods of south-west reland swamps the shrub and herb layers, and displaces holly *Ilex aquifolium* nd other native shrubs. Rhododendron's dominance is enhanced by the rampling and grazing of other plants by introduced sika deer *Cervus nippon* Usher, 1986). The trampling encourages germination of rhododendron seedings. [The taxon called *R. ponticum* in Britain is probably a mixture of hybrids of *R. ponticum* and American species such as *R. catawbiense* (Clapham *et al.*, 987; Williamson, 1994).] Vines too can swamp, and on Theodore Roosevelt sland in the Potomac River at Washington DC, Japanese honeysuckle *Lonicera japonica* inhibits the reproduction of trees, and English ivy *Hedera helix* stifles recruitment in the herb layer. Between them, they are killing the orest (Thomas, 1980). Other examples from vines such as kudzu *Pueraria lobata* are well known (US Congress OTA, 1993). Recruitment can also be set back by grazing by introduced pigs and rabbits, a more standard trophic effect.

In Chapter 5, I said the distinction between abundance effects and ecosystem effects is rather arbitrary. The distinction between abundance effects and effects on trophic structure is equally arbitrary. But trophic effects are Vitousek's third category, and many of the invaders of more serious effect, uch as cats, rats, goats, rabbits and diseases all come here.

Even generally welcome invaders, such as the sea otter (section 4.3.3) can have marked trophic effects (VanBlaricom and Estes, 1988). But the trophic effects of sea otters are complex. Fishermen dislike them because they eat fish and abalone (archaeogastropod molluscs, *Haliotis* spp., also known as ormers), but these same species are favoured by the increase in kelp beds Levin, 1988). Sea otters eat sea urchins (e.g. *Strongylocentrotus purpuratus*) hat eat kelp (several species of macro brown algae), so increasing sea otters can also (sometimes) mean increased kelp. Sea otters also eat mussels (*Mytilus californicus*), and so can lead not only to declines in species associated with mussel beds (*Halosydna insignis*, a polychaete, *Pachycheles rudis* and *Petrolisthes cinctipes*, both porcelain crabs) but also to a change of name. Mussel Point, in Monterey Bay, California, no longer notable for its mussels, s now known as Point Cabrillo (Barry *et al.*, 1995) after the navigator who sailed up the Californian coast in 1542. 'Precise prediction of the potential ecosystem level effects of an introduction of sea otters would be very difficult' Levin, 1988).

There are two reasons why I would prefer not to classify trophic effects as ecosystem effects. The first comes from defining ecosystems as biological communities in their physical and chemical environment. From that, ecosystem effects often mean changes in mineral cycles and other geochemical processes (cf. section 5.5). But some ecologists use ecosystem as a synonym for community. The other reason for not calling plain trophic effects ecosytem is that the trophic effects are still linked to a single invasive species; the knock-on effects beyond the species newly attacked are usually negligible. This is particularly so with specific biological control agents, the only desirable sort of control agents. In successful control, the target species is markedly reduced. If it is an agricultural herbivore the crop it attacks is healthier, but otherwise there are no important ecological effects. Some trophic effects, such as those caused by the brown tree snake *Boiga irregularis* (section 5.6) do go well beyond this. The situation in which trophic effects ramify so that they may justifiably be called ecosystem effects is clearly shown only in models, but may be shown also in some historic and geologic interchanges and extinctions. These two aspects are discussed in the next two sections.

7.3 COMPUTER MODELS OF ECOSYSTEM INVASION

With computer analysis and simulation, it is possible to study the properties of sets of interacting species. Such work sets standards, benchmarks, with which real communities can be compared. By being able to study a much wider range of states in a short time than is possible with real communities, it might become possible to say what patterns of invasion are normal and to be expected, what patterns are improbable or even impossible. To achieve that, computer models would have to be more realistic than has been possible up to now. Nevertheless, the results that have come from a series of intensive computer studies are interesting, and have led to searches for new types of data.

Much computer work follows, and intertwines, two threads. One starts with Elton's (1958) claim that diversity produces stability, and May's (1973) analysis of that proposition. The other comes from the empirical and theoretical study of food-webs (Pimm, 1982; Cohen, Briand and Newman, 1990). From the conclusion that large randomly constructed theoretical communities are unstable, one line followed was to see how large a stable community could be constructed by introducing species successively, by using models of invasions. Almost all this type of work has been done using what are known as trophic species in Lotka–Volterra equations. Those equations are described in the box.

Trophic species are theoretical entities defined by what they eat and what eats them; their food links are invariant. Real species are by no means always trophic species. For instance, frogs and tadpoles are part of the same species but very different trophically, and many species have different trophic relations at different stages of their life history. Food habits may change seasonally or be dependent on availability in a way not followed by the models. Williams (1992) had other criticisms. The point is that the use of trophic species is but one rather obvious way in which computer models are, as yet, rather distorted images of reality.

Similarly, it would be possible to use rather more realistic mathematical models than Lotka–Volterra models, such as those used in epidemiology (Anderson, 1994) or for continuous cultures (Williamson, 1989b). As so often, increased reality would mean less generality. For small numbers of species quite general models can be studied (Hutson and Law, 1985), but for larger numbers, and particularly when studying the mathematical concept of permanence, discussed below, technical difficulties mean that Lotka–Volterra models are studied rather than any others. In almost all studies, the Lotka–Volterra communities have been either food-web models or competition models. In food-web models, the species are defined trophically, and all links are vertical, with flow from the basal producers to the top species which are generally called top predators. In competition communities, all the links are horizontal, between members of the same trophic layer. Food-web models regard all competition as either going down and up, through resources, or (apparent competition) going up and then down through a predator. Competition communities ignore all other trophic levels, a tradition that is familiar, say, in much terrestrial plant ecology.

Virtually all Lotka–Volterra systems have a single point of equilibrium (see box) away from the axes. If that point is a stable and feasible equilibrium, then it represents a stable and persistent community. Stable is a word with many meanings in ecology (Grimm, Schmidt and Wissel, 1992), but in these model systems a stable point is simply one the system will move to and stay at. Feasible in this context means that no species has negative numbers, or, to put it technically, that the point is in the positive orthant (see box).

When the equilibrium is feasible but unstable, or when it is not feasible whether stable or not, then the situation is more complicated. Post and Pimm (1983) and Drake (1990) used a variety of heuristic (plausible) rules to decide what would happen in food-web models. In some circumstances some of those rules were misleading, but more seriously the study of local stability is probably not what is needed (Morton *et al.*, in press). Ecologists want to know not whether a community will be rock steady, but whether it will persist. In mathematical terms, the concept of permanence comes close to what is wanted. In a mathematically permanent system, all the boundaries of the positive orthant are repellors (see box). That is, the trajectory of the community in multi-dimensional space will move away from a boundary if it is close to one. More intelligibly, that means that if any species is rare, at almost zero density, it will increase.

The mathematics of permanence are complex (Hofbauer and Sigmund, 1988; Hutson and Schmitt, 1992), but it is possible to test for permanence in most Lotka–Volterra systems. The cases where the issue is in doubt are sufficiently few that their community trajectories, the successive population sizes of all species, can be calculated (Law and Morton, 1993). For a set of competition communities, chosen randomly and not constructed by

Lotka–Volterra equations

This box is a summary of the properties of these equations, and some terms used in population dynamics. It is not intended as a text on the subject.

Lotka–Volterra equations were developed independently in the first quarter of the 20th century by Lotka, an American, and Volterra, an Italian. They are among the simplest that could be devised to describe the population dynamics of sets of species. Mathematically, they are relatively easy to handle because of the linear part of them, explained below. The basic equation for one species is the logistic equation, first used in the 19th century. Ecologists usually write this

$$dn/dt = rn\,\frac{(K - n)}{K}, \qquad [1]$$

where n is the population size, which has to be non-negative to have biological meaning, t is time, r is the intrinsic rate of increase and K is the equilibrium population size usually called, to my mind somewhat misleadingly, the carrying capacity. To get this in a form suitable for many species, it can be rewritten as

$$d\ln n_1/dt = r_1 + a_{1,1}n_1, \qquad [2]$$

where $a_{1,1}$ is a parameter describing the density-dependent effect of the species on itself. $a_{1,1} = -r_1/K_1$. The subscripts have been added to show that all the terms refer to species one. For several reasons the parameter a is to be preferred to the parameter K, one being that it allows an easier notation for equations describing more than one species. r is also a parameter, so both equations [1] and [2] have two parameters.

For two species the corresponding equations are

$$\begin{aligned} d\ln n_1/dt &= r_1 + a_{1,1}n_1 + a_{1,2}n_2 \\ d\ln n_2/dt &= r_2 + a_{2,1}n_1 + a_{2,2}n_2. \end{aligned} \qquad [3]$$

If the r_i are positive and all the a_{ij} are negative, then the equations represent competition, with each species having a negative density dependent effect on itself and the other one. However, with r_2 negative and $a_{2,1}$ positive, the equations represent a predator–prey system. The original Volterra predator–prey equations also had $a_{1,1} = a_{2,2} = 0$, giving only four terms on the right-hand side. With r_i negative, r no longer refers to the intrinsic rate of increase of the predator (Williamson, 1989a), but to its rate of decrease under starvation.

Equations [3] can be written in a form that can relate to any number of species

$$d\ln n_i/dt = r_i + \Sigma_j\, a_{ij}n_j \qquad [4]$$

The right-hand side of equation [4] is linear, but the variable on the left is ln n not n, so the set of equations are not linear. However, if you set the left-hand side equal to zero, that gives the locus of equilibrium for each species, and the system becomes

$$r_i = -\Sigma_j a_{ij} n_j \qquad [5]$$

and that is now a linear system. The equilibrium locus for each species is a hyper-plane, i.e. a straight line with two species, a flat plane with three and so on. With such a linear system, provided none of the hyper-planes are parallel, mathematically there is one and only one equilibrium point. It may be a stable or unstable equilibrium; it may be biologically impossible (not feasible) by having one or more of the n_i negative. A stable equilibrium is an attractor, trajectories move towards it. An unstable equilibrium is a repellor, trajectories move away from it. Trajectories are the paths taken by the point representing the abundance of the set of species.

However, the biological requirement of feasibility means that all the intersections of the hyper-planes and the axes are also (biological) equilibria of the system. This requirement can, of course, be put in mathematical form with the extra inequalities

$$n_i \geq 0 \qquad [6]$$

Another way of saying the same thing is to say that the equations only have biological meaning in the positive orthant. An orthant is a segment defined by the axes and the origin. So the positive orthant of a line is the positive half, the half to the right of, and including, zero in the usual convention. In two dimensions, the positive orthant is the positive quadrant, the upper right-hand quadrant, the sector of the plane where both x and y axes are positive.

Lotka–Volterra equations are popular with theoreticians, as they are reasonably tractable mathematically, and show a range of dynamics. However, Hirsch and Smale (1974) showed that a much wider range of dynamical outcomes is possible in more non-linear two-species systems. Multiple equilibria are only possible in a non-linear system, and even a two-species non-linear system can have more than one equilibrium point.

Outcomes can vary drastically with slightly different initial conditions. 'The moral is clear: in the absence of comprehensive knowledge, a deliberate change in the ecology, even an apparently minor one, is a very risky proposition.' (Hirsch and Smale, 1974, p. 273).

successive invasions, Case (1990) calculated the trajectory at each invasion.

Mathematical permanence is not a perfect ecological concept. Almost zero in a mathematical sense might imply really zero for a species of discrete individuals. In some models there could be a stretch of the boundary, the boundary where one or more species becomes zero, that was an attractor rather than a repellor but nevertheless the bulk of the system maintains the species, the bulk is permanent in the colloquial sense. Given the limitations of computer models discussed earlier, the limitations of the concept of permanence are rather trivial.

So what conclusions have been drawn? I will pick out five: size, non-equilibrium, transients, alternative endpoints; and resistance to invasion. First, on size, all simulations have ended with a small number of species, typically around ten. This is so whether the pool of available invaders is fairly small, rather large or theoretically infinite. The results are, of course, for isolated homogeneous communities and for a small set of models, but may indicate the importance of heterogeneity, dispersal, variation in life history, and other factors not in the models, in maintaining the larger number of species seen in almost all real communities. For invasions, the message may well be that it is more difficult to add species than the structure of the community may suggest.

Second, if there is an unstable feasible equilibrium, then the system may still be permanent. The trajectory will either be a limit cycle or a chaotic one. Both are non-equilibrial, though stable in some senses of that word (Williamson, 1987). Law and Blackford (1992) found such non-equilibrial communities quite common, 40% or more of cases, in food-web models with omnivore links. An omnivore, in the modelling sense, is a species that feeds on species in more than one trophic layer. A secondary predator that feeds on a herbivore as well as a primary predator is an example. Proven examples of limit cycles or chaos in natural populations are rare, but this may be partly because there are very few sets of good long-term data. For invasions, the fact that a community is variable does not make it more invasible, or less permanent.

A transient species, the third point, is, in the sense used here, one that invades and then dies out, a boom-and-bust species (section 2.4.1). In models, that means dies out before the next invader arrives, because nearly all models have studied invaders one at a time. Law and Morton (in press) find no transient species in their food-web communities. Any species that manages to invade becomes part of a permanent set, at least until the next invasion. Case (1990), with competition communities, distinguished rejection failure, where the invader is repelled, never gets going, from indirect failure, where the invader grows but later dies out. In indirect failure the invader fractures the community, and one or more resident species die out with the invader. Unfortunately, he combines the two sorts of failure in presenting his results, but he implies that indirect failures are at least not rare.

Transients are known from real data, for instance in introduced birds in Hawaii (section 2.3.2), but have never been associated with the loss of other species a result intermediate between the Case and the Law and Morton conclusions. More data on the frequency of transients and on their effects might at least be useful in developing models.

Alternative states are the fourth point. A permanent state implies that a particular set of species coexist indefinitely, though their populations may be variable. Models have been known for some time (Hofbauer and Sigmund, 1988) in which no set is permanent though the community as a whole persists. With two predators and two herbivores, and using an obvious notation, it is possible to find heteroclinal systems, where each successive invasion leads to a new set of coexisting species, such as:

$$H_1 \rightarrow P_1H_1 \rightarrow H_2 \rightarrow P_2H_2 \rightarrow H_1 \ldots$$

or:

$$P_1H_1 \rightarrow P_1H_2 \rightarrow P_2H_2 \rightarrow P_2H_1 \rightarrow P_1H_1 \ldots$$

As well as such cases, which have no permanent state, Law and Morton (1993) found alternative permanent states, where the community could go to more than one non-invasible set. For instance, in a set of five species, two basal (A and B), two intermediate (C and D) and one top (E) they found, with a suitable choice of fixed parameters, {B, D}, {A, C, D} and {A, C, E} to be permanent and uninvasible by any of the other species. It is difficult to think of invasion situations in which the replication would allow such phenomena to be studied, but they could be important when trophic interactions are strong.

The fifth and last point is resistance to invasion. In most models resistance to invasion develops as the model is run, it becomes progressively harder to find an invader that will succeed. Law and Morton (in press) found that resistance could take some time to build up, more so with more species in the pool, but the eventual outcome was always either permanent and uninvasible, or heteroclinal. Invasion either became impossible or restricted to very particular species. Case (1990), working with a notionally infinite set, found that invasion resistance was more marked the stronger the interaction. As real communities seem always to be invasible (CFP 3, section 3.2), this is perhaps a reminder of the remarkable variety of natural species and an indication that most interactions between species in real communities are weak. The strength of interactions is an important question I will return to in section 7.4.4 below.

Both the fourth and fifth points suggest that invaders change communities in ways that allow other invaders in. With successive invasions, and after some individual invasions, there is a cascade of effects that alter the composition and balance of the community. In the examples below, we can search for such cascades.

In a rather different sort of study, Yodzis (1988) built models to imitate some real communities, 16 communities drawn from the literature (Cohen, Briand and Newman, 1990) with from 3 to 32 (trophic) species. He studied the effect of press perturbations, that is, of continually adding or removing individuals of one species. So his equations were a slightly modified form of Lotka–Volterra predator–prey equations. The effects can be classified as direct when they affect the species directly linked, feeding on or being fed on the perturbed species, indirect otherwise. Because of indirect effects the results were highly indeterminate. His model ecosystems acted as complex systems in which an effect at one point would affect many other points in ways which were very difficult to understand, let alone predict.

A successful invasion is a permanent (i.e. a press) perturbation of an ecosystem. In Yodzis' models, using fairly strong interaction terms, the consequence of indirect effects is that the eventual effect of the invader on the community appears indeterminate. Is this a conclusion about models or about real communities? This is a question for section 7.4.4, after the discussion of some more invaded communities.

7.4 INTERCHANGES, EXTINCTIONS AND COMMUNITIES

Computer models suggest that there should be, sometimes, cascading or chain effects of invasions. The introduction of species A affects B (and others), the effect on B affects C which affects D and so on. But the examples in section 7.2 suggest that these indirect effects are either overlooked or heavily damped. The major invader may affect many species, by attacking them or by changing the physicochemical regime or in other ways, but the effects of those on yet other species are usually not discussed. Possibly the task is too big for a normal research team, possibly the effects take too long to develop. But it may be that the normal structure of ecological communities is such that such effects are usually too small to be demonstrable. If so, the study of invasions is important for understanding the structure of communities (and how to model them) and for the conservation of biodiversity.

Most of the examples of invasions show little knock-on effect. Back in Chapter 1, *Veronica filiformis* (section 1.1) and fulmar (section 1.3.1) have had negligible effects on any other species while becoming quite common. Clearly they must have had some effect; where *V. filiformis* grows thickly it must have displaced other plants. With disasters such as *Boiga* and *Euglandina* in Chapter 5, or with biological control agents such as *Chrysolina quadrigemina* controlling *Hypericum perforatum*, the species attacked have gone extinct or declined, an A → B reaction rather than an A only effect, but there is little recorded A → B → C reaction. A biological control agent for an insect pest, where the agent A controls the pest B leading to a healthier crop C, is about as far as cascades have gone in the examples considered.

7.4.1 The Flathead catchment system

There is one case reported in which indirect effects are stressed. The scale both in space and time may be one reason why such effects have not been reported. The case is the Flathead River–Lake catchment in the north-west of the state of Montana, USA, overlapping part of Glacier National Park (Spencer, McClelland and Stanford, 1991).

Managers of freshwater ecosystems have been remarkably willing to try introductions, particularly to improve fisheries. Usually, they import new species of fish; sometimes they import other sorts of species to increase the fish stock. The statistical consequences of these actions have been considered in section 5.1. The Flathead catchment is 22 241 km^2, with two major hydroelectric dams completed in 1938 and 1952 and six lakes. There are 10 native species of fish; 17 species have been introduced, 14 of them between 1898 and 1916, though in some cases there were multiple introductions after that. Two of the introductions are now absent, nine have a limited range, four a moderate one and two have extensive ranges. Kokanee (or landlocked sockeye salmon) *Oncorhynchus nerka* was introduced in 1916 and replaced westslope cutthroat trout *O. clarki lewisi* as the dominant sport fish. Kokanee became a major food source for many species, as we shall see.

Mysis relicta (Figure 7.1) the freshwater opossum shrimp is a remarkable species. Mysids are a mostly marine group that get their English name from carrying their young in a brood pouch. *Mysis oculata* is an Arctic Ocean species. The classical theory is that as *M. oculata* spread north after the last glaciation it gave rise to populations of *M. relicta* in freshwater lakes in Scandinavia, Germany, the British Isles and North America, including the Great Lakes, as well as in some brackish waters such as the northern Baltic (Tattersall and Tattersall, 1951). If so, it seems *M. relicta* arose several times, but there are other theories (Gledhill *et al.*, 1976).

In 1949, it was found that introducing *M. relicta* had a dramatic effect on the growth of kokanee in Kootenay Lake, British Columbia. This led to it being introduced to more than a hundred other lakes in western America, including the upper Flathead catchment between 1968 and 1975. They reached Flathead Lake, the largest in the catchment, in 1981. This produced marked changes in many populations, notably decreases in most cladocerans and copepods. Two cladocerans, *Daphnia longiremis*, a herbivore, and *Leptodora kindtii*, a carnivore, disappeared from samples in the early 1980s, to reappear at low densities in 1988 and 1989. Kokanee, at least in this catchment, fed on the copepods and cladocera rather than on *Mysis* and its population crashed in 1986. The count of spawners in the catchment varied from 26 000 to 118 000 between 1979 and 1985, and fell to 330 in 1987, 50 in 1989, a 500 : 1 reduction.

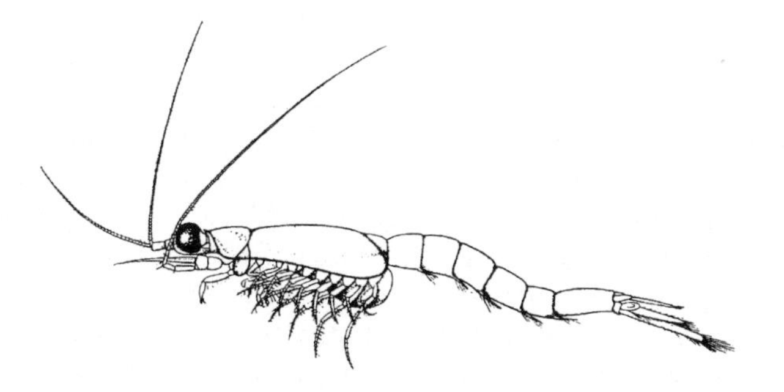

Figure 7.1 Mysis relicta, female, side view, showing the pouch. The length of adults of both sexes is 15–18 mm. (After Sars, 1867.)

Before the crash, McDonald Creek in Glacier National Park attracted a spectacular diversity and abundance of birds and mammals each autumn, feeding on spawning kokanee, carcasses and eggs. The hundreds of bald eagles, *Haliaeetus leucocephalus*, were a particular tourist attraction. Only 25 were seen in 1989. Other species affected, at least locally and in the autumn, were two gulls, herring gull *Larus argentatus* and California gull *L. californicus*, four ducks, merganser *Mergus merganser*, mallard *Anas platyrhynchos*, goldeneye *Bucephala clangula*, and Barrow's goldeneye *B. islandica* and also American dipper *Cinclus mexicana*. Mammals involved included four carnivores, grizzly bear *Ursus arctos*, coyote *Canis latrans*, mink *Mustela vison* and otter *Lutra canadensis* and, more surprisingly, whitetail deer *Odocoileus virginianus*. Tourists declined from 46 500 in 1983 to less than 1000 in 1989. Naturally no tourist died as a result, and similarly it is not known what effect there was on the population sizes of the birds and mammals listed.

Spencer, McClelland and Stanford (1991) speculate on the likely population effects on the birds and mammals and suggesting that bottom-feeding fish such as lake whitefish *Coregonus clupeaformis* and lake trout *Salvelinus namaycush* (both introduced) may gain long term. But the strength of the indirect effects is not clear, though their direction seems thoroughly predictable. The Flathead Lake system gives only weak support to Yodzis' contention that unpredictable indirect effects will be the typical consequence of a press perturbation. Still this is a possible cascade of *Mysis* to kokanee to various birds and mammals to other fish, an A → B → C → D cascade.

7.4.2 Interchanges

As so many species have been introduced into the Flathead catchment, the system there has some features of an interchange. The evolutionary effects of interchanges, when two biotas meet over a geologically long time, were discussed in section 6.4.3. Vermeij (1991a) also considered the ecological effect of 12 interchanges. His major conclusion was that relatively impoverished biotas are more susceptible to invasion, and that impoverishment often comes from extinctions before the interchange. So interchanges are often asymmetric. His views fit with the traditional view that invasions are aided by empty niches. The difficulties in that were discussed in section 3.4.6. Tautologously, richer biotas have more species, and so a greater ecological variety of species. That by itself increases their chance of producing a successful invader.

The Great American Biotic Interchange (Stehli and Webb, 1985) shows the difficulties of drawing ecological conclusions from these ancient and long-lasting events. At the level of mammalian families, as used by Simpson in his classical studies, there are remarkable examples of convergence, in appearance and ecology, of North and South American forms. Some are illustrated in Wilson (1992). South American carnivores died out, North American carnivores succeeded them. It is not clear what is the relative importance of competition, habitat change and the normal turnover of origination and extinction seen in the geological record. At the generic level, there was about a 25% loss of North American genera in North America and of South American genera in South America during the Pliocene (Marshall *et al.*, 1982), then a general increase of genera during the Pleistocene before the major extinctions at the end of the Pleistocene and in the Holocene.

While there is still much taxonomic and timing detail to be unravelled, the importance of habitat, and so of climate change, is becoming clear. Broadly, tropical forest taxa of plants, birds and mammals have moved predominantly from south to north, savanna taxa from north to south (Vermeij, 1991a; Webb, 1991). Those generalizations fit with the relative areas of the biomes in the two continents. A larger area implies more species, and the imbalance in the numbers of species may possibly explain the main features of the interchange. Over several million years there is the added complication of speciation and evolution. North American savanna mammals produced a remarkable number of descendent taxa in South America (Webb, 1991). There is nothing in the current view of this interchange that suggests a cascade effect from invasion.

The same seems to be true of the one modern interchange discussed by Vermeij (1991a), the effect of building the Suez Canal on the ecosystems of the eastern Mediterranean and the Red Sea. Such a recent interchange means that it is much easier to be sure of the species and timing involved. It is also a nice example of how details of the physical environment may influence the timing and outcome of events.

The 160-km long Suez canal was completed in 1869 under the direction of Ferdinand de Lesseps, so Por (1978, 1990) has called these invasions Lessepsian. The southern part of the Canal opens into the Gulf of Suez, a horn at the north end of the Red Sea. The salinity throughout the Red Sea is high, with a marked increase in the north and reaching 43‰. The Canal runs through the Great and Little Bitter Lakes, which had salinities when the Canal was completed of 161‰ (Galil, 1993), though they are down to 45–47‰ now. The eastern Mediterranean, at the north end of the Canal, is only slightly saltier than the open ocean, 35–37‰. But before the Aswan Dam was built on the Nile in 1967, the Nile floods led seasonally to salinities as low as 26‰ at the north end. That has now ceased. There has been a complicated and variable salinity barrier to the interchange, the barrier decreasing over the years.

There may also be a temperature barrier, particularly a winter temperature problem for species going north. The current will have influenced migration more strongly. From October to June the Canal flows north at up to 2 knots. In the other three months the current is usually to the south, but evaporation at the Bitter lakes may make it flow from both ends towards the middle.

The number of species that have migrated is far from clear, but is still increasing. The Red Sea is the western appendix of the rich Indo-Pacific province, the Mediterranean the eastern appendix of the rather poorer Atlantic flora and fauna. In part this comes from the destruction of the Tethys

Ocean by tectonic movements (the formation of the Himalayas, etc.) in the Tertiary, which led to differential extinction (Vermeij, 1991a). However, there are quite a few pan-tropical and other species that probably occurred at both ends of the Canal before it was built. Por (1978) listed only those going north, and counts as 128 high probability Lessepsian migrants, 76 as low probability ones, 204 in all. Of these, 119 had been found in the Suez Canal. Taking only the high probability species matches Safriel and Ritte (1986) who said there are about 130 Red Sea-to-Mediterranean invaders, but only about 10–15 the other way, a marked imbalance. The invaders include diatoms, macrophytic algae, sponges, a pycnogonid, copepods, decapods and other crustacea, a chiton, bivalves, gastropods, bryozoans, echinoderms, tunicates and fish. Many have planktonic larvae.

Lessepsian invasions conform to many of the earlier conclusions. First propagule pressure or abundance in the native habitat may be important Golani (1993) found that abundant species on sandy shores in the Gulf o Aqaba (the other horn at the northern end of the Red Sea) were more likely to be successful Red Sea migrants than scarcer ones, but the relationship is weak though statistically significant. Second, the idea that it is *r*-selected species tha migrate does not match some very limited data (Safriel and Ritte, 1986).

Some Lessepsian migrants have become very common in the Mediterranean A large scyphozoan jellyfish, typically 50 cm in diameter, with symbiotic photo-synthetic zooxanthellae, *Rhopilema nomadica*, was first found in 1977 but now occurs in aggregations of between 1 and 10 per m^3; 130 can be stranded per 100 m of beach. *Saurida undosquamis*, the brushtooth lizardfish, makes up to 20% o trawl catches; *Upeneus moluccensis*, a goatfish, one-third of the mullet catches *Sphyraena chrysotaenia*, half the barracuda catch (Galil, 1993). While some ma have become more abundant in the Mediterranean, the data are again consisten with the view that more abundant species are more likely to be successfu invaders (section 3.4.3). *U. moluccensis* is one of several species which hav shown a boom-and-bust (section 2.4.1), going from slightly more than 10% t over 80% and back to around 30% of the mullet catches in a period of decade (Galil, 1993).

There is some evidence of competitive displacement, and sharing out o habitats (Galil, 1993). A native Mediterranean penaid prawn, *Penaeu kerathurus*, supported a commercial fishery in the 1950s, but it is now scarc and Red Sea-derived prawns common. The immigrant *Alpheus inopinatus*, snapping shrimp, is now more common than the native *A. dentipes* in the rock infralittoral (the highest zone below the shore line). *A. glaber*, a native, is foun from 45–145 m depth while the immigrant *A. rapacida* occurs between 15 an 50 m, having apparently displaced *A. glaber*. *Rhizostoma pulmo*, a nati jellyfish in the same family as *Rhopilema nomadica*, though without photo-synthetic symbionts, has become rare. All in all, Lessepsian migrants follo the pattern of other invasions, the majority without obvious effects on oth species, occasional cases of possible (though unproved) stronger interaction but no community-wide patterns or cascades.

7.4.3 Quaternary extinctions

Community-wide patterns and cascades are perhaps to be found in the record of quaternary extinctions. The quaternary is the geological era covering the Pleistocene or ice-age and the Holocene or recent. The Pleistocene started 1.64 million years ago (Harland *et al.*, 1990) and the boundary between it and the Holocene is put at 0.01 million years ago, or 10 000 radiocarbon years before 1950 AD.

There are two major sets of extinctions in the quaternary that may show cascades of extinction following an invasion, but there are complications of habitat and climate change that make interpretation difficult. The two sets are the extinctions of mammalian megafauna, particularly in the Americas, at the end of the Pleistocene and the extinction of birds on islands in the Pacific. In both cases the primary invader is that most destructive of all species, *Homo sapiens*.

The human role is clearest with the birds, so consider them first. Where humans, particularly Polynesians, settled on islands in the Pacific many birds became extinct. Some of these have been known for some time, most strikingly the giant flightless moas of New Zealand. Anderson (1990) considers that there were 13 species, and that reliable radiocarbon dates for hunting sites date from about 1000 bp (= 950 AD), ending at 300–400 bp (ca. 1600 AD). [Others now count only 11 species (Cooper *et al.*, 1993), but the exact number does not affect the argument here.] The Maoris not only hunted moas but also changed some habitats by agriculture and burning.

Many other New Zealand birds went extinct in this period. Cassels (1984) listed 15 other endemic species, two subspecies and four species that lost their New Zealand populations. Taxonomic revision and new discoveries will affect the number, but the list includes several rails, a swan, geese and ducks, a pelican, a crow, an eagle and other hawks, and an owlet-nightjar. One of the extinct ducks was the New Zealand population of *Oxyura australis* (see Figure 6.4). While some of these may have been hunted directly, others such as the birds of prey may have gone extinct from lack of food. Others may have lost their habitat. It is possible that there is a cascade here

A (*Homo*) → B (moa) → C (predator) → D (other prey species),

if the predators had to concentrate on other species before becoming extinct, but that is very speculative. Cassels (1984) points out that the extinct birds were often large, flightless and diurnal, living in the forest ecotone. Flightless nocturnal birds of the primary forest, notably the three species of kiwis *Apteryx*, survived.

Excavation have now produced similar stories on many oceanic Pacific islands. Steadman (1995) thinks 1600 bird species may have disappeared within a few hundred years of settlement, an enormous number in relation to the 9000 or so bird species extant. On Hawaii for instance (James and Olson, 1991; Olson and James, 1991) at least 50 species went extinct, including two flightless ibises, four flightless goose-like ducks, six hawks and owls, various rails and crows and a couple of dozen of that most distinctive group the Hawaiian Drepanidines or honey-creepers. Curiously, the finch-like and insectivorous honey-creepers seem to have been more susceptible than the nectar-feeding ones. The Polynesians of Hawaii practised agriculture, and brought dogs, pigs and the Polynesian rat with them, so there are many possible causes for these extinctions.

While the direct and indirect role of humans in those bird extinctions is clear, there is much more dispute about what happened in megafaunal extinctions. Towards the end of the Pleistocene, as is well known, many large mammals went extinct. Examples are mammoths and mastodons, woolly rhinoceros, Irish elk, sabretooth cats, giant sloths and various giant marsupials in Australia. A comprehensive bestiary is given by Anderson (1984).

The timing and the degree of extinction was different in different places, the peak of extinction being earlier in Australia (26–20 kyr bp) compared with the Americas and Europe (12–10 kyr bp). Taking the whole period of the last glaciation, from 130 kyr bp, and a very broad definition of megafauna as more than 5 kg body weight, Owen-Smith (1987) found the percentage loss of genera to be 75% in North America, 76% in South America, dropping to 45% in Europe and 43% in Australia and only 13.5% in Africa. The effect by size is just as striking. Excluding Africa (and Asia), but again taking percentage extinctions of genera, Owen-Smith found only 1.3% in mammals less than 5 kg (the normal sort of background extinction), 41% in those 5–100 kg, 76% for 100–1000 kg and complete extinction, 100%, for those over a metric ton, meaning proboscideans, rhinos and such-like huge animals.

In a famous symposium (Martin and Klein, 1984) the contributors tended to ascribe these late Pleistocene megafaunal extinctions either to climate change or to hunting by man. Neither explanation is very convincing as a sole and single cause. The climate did change very rapidly at the end of the Pleistocene, but then so it had at the end of earlier glaciations without such a marked and biased set of extinctions. Similarly, human hunters were found for the first time at the end of a glaciation outside Africa or Asia, but for most of the extinct species there is no sign of hunting. Only Diamond (1984) suggests a trophic cascade, though his modern examples are only two- or three-step (AB or ABC) ones, such as the need for the tree *Calvaria major*, still surviving on Mauritius, to have its seeds eaten by dodos *Raphus cucullatus* before they can germinate (Temple, 1977).

The problem of extinction in the absence of obvious hunting is brought out clearly in Guilday's (1984) analysis of late Pleistocene and Holocene mammals in the central Appalachians in eastern USA. There, only bison, mammoth and mastodon were known to have been hunted, though horse and muskox may have been. But the extinct species includes two sloths and an armadillo, a giant beaver, two peccaries, two deer, a tapir, a skunk, a bear or maybe two, a sabretooth and possibly a cheetah, and the dire wolf.

A cascade from the loss of mammoths to switching by the predators to the loss of other herbivores and eventually the loss of the large carnivores is possible. Owen-Smith (1987, 1988) suggests a complementary hypothesis. Mega-herbivores such as elephants have a major role in maintaining a heterogeneous environment. With their loss from hunting, habitats would change, facilitating if not driving the extinction of other species. Both hypotheses, or a mixture, imply drastic community effects from the arrival of a single new species, *Homo sapiens*.

Quaternary extinctions do suggest that community-wide changes can result from single invasions, but such changes seem to be highly exceptional.

7.4.4 Community conclusions: value of richness, alternative states, strength of interactions

It is now time to look at the relevance for the effect of communities on invasions at three points that have so far just been touched on. First, are some communities more invasible? Second, can communities have alternative states and, if so, can invasive species switch the community from one state to another? Thirdly, and arising to some extent from the first two, how strong are the links in communities, and what are the implications of that for the success and consequences of invading species?

First, the invasibility of communities. This was one of the questions in the SCOPE programme (see Chapter 1, Appendix) and on the whole not answered by it. It is also a question raised by Elton (1958), who thought diverse communities less invasible, confronted by May (1973) who found no good theoretical reasons why Elton should be right, and the subject of innumerable studies thereafter. In the computer models discussed above (section 7.3), communities do become less invasible after repeated invasions, but this has not been related to any diversity measure. Other theoretical studies, summarized by Roughgarden (1989), lead as yet to no simple conclusions about the invasibility of communities, but in a few individual cases the relation of the structure of the community to the pool of possible invaders and to species interactions is starting to be understood. Many more such cases are likely to be needed to give valid generalizations.

Still, it is possible to say something empirically about Elton's generalization, and that is that it is scarcely, if at all, supported by fuller data. Several contributors to the SCOPE volume on tropical invasions (Ramakrishnan, 1991) conclude that many tropical systems are invasible, and quite possibly as invasible as temperate systems, though it is difficult to make good quantitative comparisons. The water fern *Salvinia* (section 5.4) and the Nile perch *Lates niloticus* (section 5.2.2) are obvious examples of major invasions that were entirely tropical. One was into a species-poor system, one into a species-rich one.

Rejmánek (in press) has now made a thorough quantitative study of invasibility by plants in relation to latitude, and comparing islands and continents. On islands, he finds no significant variation of invasibility with latitude. One of his measures is the percentage of alien species. His tropic data have a very large variance, with the highest and among the lowe percentages, a range from 5% to 90%. In contrast, the continental data do ha a clear, almost U-shaped, pattern in the percent aliens, from a peak of abo 20% near 40°S, through a trough of around 5% with little variance in t tropics, rising to a higher peak of around 35% at 45°N, but then falling aw to 5–15% at 60°N. South of 40°S the only continental land, apart fro Antarctica, is the thin end of South America. 45°N runs from Portlan Oregon to Halifax, Nova Scotia, through Bordeaux, France, the Crimea ar on to Japan's northern main island of Hokkaido.

As Rejmánek points out, his results give little support to the idea th richness *per se* is important in determining invasion. For instance, the remar ably species-rich fynbos in South Africa (Cowling, 1992) is also remarkab susceptible to invasion. So, although different ecosystems do indeed ha different records of invasion success, it is not yet possible to say why this is s Propagule pressure (section 2.4.2) accounts for some only of the variation.

The second topic of this section is the question of alternative states. If community has alternative states, either different sets of densities at which t same species are in equilibrium, or different sets of species in a state (mathematical) permanence (as discussed above in section 7.3), then t community might be harder to invade than communities that have only single equilibrium. The reason for thinking that that might be so is that t conditions for a successful invasion are likely to be more stringent, mo restricted, when there are multiple states. Unfortunately, the field evidence f multiple states, let alone the effect of such states on invasibility, is thin.

Multiple equilibria with the same set of species are impossible in Lotk Volterra systems, but easy to model if non-linear isoclines are allow (Williamson, 1957; Hirsch and Smale, 1974). In eastern Canada *Choristoneu fumeriferana* spruce budworm is a major pest of coniferous forests. It attac particularly *Abies balsamea* balsam fir, but also various *Picea* spp., spruce Outbreaks retard growth and can kill trees. There were widespread outbrea in the 1950s and the 1980s, and the population was scarce in-between. T interpretation in the 1960s was that there was an outbreak and a non-outbre equilibrium, with the system switching at fairly long intervals. Royama (199 has shown that, although the system is still far from fully understood, altern tive equilibria are not involved.

A reasonably well-documented case of alternative states is the communi of the sub-littoral on two islands off South Africa reported by Barkai ar McQuaid (1988). The two species are *Burnupenia papyracea*, a whelk and *Jas lalandii*, a rock lobster. Each as an adult preys on the young of the other. S where the adult of one is well established, the other is prevented from invadin leading to states dominated by either one or the other. These states we observed on neighbouring islands, *Burnupenia* dominating on Marcus, *Jas* on Malgas. It will be interesting to see if they are as permanent as Barkai ar McQuaid thought. Their observations covered 4 years. The example was us by Williams (1992) to point to the fallacy of equating niches with species or vice-versa.

Shallow freshwater lakes may show alternative states more commonly (Scheffer *et al.*, 1993). These are lakes only a few metres deep, and there seem to be two common states, turbid and clear. In a clear lake the bottom is covered with plants and plankton is scarce. In a turbid lake there is little bottom vegetation and much plankton and other suspended matter. The turbid state is stable because bottom vegetation cannot grow for lack of light. There are four reasons for the stability of clear lakes. The plants minimize resuspension of silt and mud; the plants are a refuge from fish for zooplankton, which reduces the phytoplankton; the plants sequester nutrients which slows algal growth; and the plants release allelopathics toxic to phytoplankton. At low nutrient concentrations lakes are clear, at high they are turbid. Both states are possible at intermediate nutrient concentrations, and there is evidence that some lakes may switch, or may be managed so as to switch, from one state to the other. Which species could invade would presumably be different in the two states. Alternative states may be important, and may explain some of the variability in invasion success.

After two rather negative results of trying to tie invasions to community structure, the third, on the strength of interactions is rather more positive. Modellers have often assumed that species interactions within a community are strong and fixed. Evidence is growing that many interactions are in fact rather weak and variable. Much natural history would support that; many species are opportunistic feeders, changing their food preferences depending on the densities of different prey species. There are, of course, specialists and for such species interactions with their particular prey are stronger and less, if at all, labile.

The evidence that interactions are weak is itself weak. Paine (1992) experimented to find the strength of interactions of a guild of littoral herbivores with the algae they fed on. He used one sea urchin species, three chiton species and three species of limpet. Mostly the effects were feeble, though *Strongylocentrotus purpuratus* the sea urchin and *Katharina tunicata*, one of the chitons, had strong negative effects which were not additive, as *Strongylocentrotus* excludes *Katharina.* Some of the weak effects were positive, because of preferential grazing on encrusting species on the target algae. Lawton (1992) points out that many ecologists study only single interactions, which produces an inevitable bias to studying strong ones.

In a more wide-ranging review, but heavily dependent on Paine's and their own results, Hall and Raffaelli (1993) suggest that the evidence from a range of field experiments suggests that most interactions are weak, though a small proportion are strong, and contrast this with the assumptions used by a variety of modellers. However, partly because of this work, Law and Morton (1993 and in press) used a negative exponential distribution for their interaction terms, the a_{ij}s (see box). Their conclusions were discussed above.

There have also been several papers, notably in the *American Naturalist*

(Winemiller, 1989; Warren, 1990; Polis, 1991; Martinez, 1992) showing th
when ecologists look systematically for all interactions in a food-web they fir
many more, and more variable connections than has been usually bee
reported in the literature so ably summarized by Cohen, Briand and Newma
(1990).

If weak and variable interactions are the norm in ecological communitie
how does this fit with what is known of invasions? It is certainly consiste
with most successful invasive species having minor or negligible effects, ar
the low frequency of strong effects is consistent with about 10% becomir
pests. The low frequency of successful invaders would seem to be unrelate
being perhaps more a consequence of inadequate propagule pressure and
variety of particular requirements that are not met in the community bei
invaded. Certainly where several species invade, as for instance in the pos
glacial spread of trees (section 4.4) each species appears to spread indepen
ently of the others. Similarly, the scarcity of identifiable community effect
such as cascades, discussed in this chapter, are consistent with generally wea
interactions in most communities.

7.5 CONCLUSION

Invasions are important, invasions are interesting, invasions are inadequate
understood. Those conclusions apply to all stages of invasions, while th
structure of our understanding of invasions is set out in the CFPs in Table 1
and the chapter heads. Some aspects of some invasions are well understoo
but for the majority of invasions, ecology can only, as yet, strive to impro
the statistical and approximate conclusions such as the tens rule of Chapter
or the results from the study of spread in Chapter 4. When deliberate intr
ductions have a marked effect, these are usually mostly undesirable. A few a
desirable, notably the success stories of classical biological control. But the
successes have to be put against disasters like the effect of *Euglandina rosea* c
endemic and irreplaceable snail species throughout the Pacific (section 5.6
Biological control in future should be absolutely specific and guaranteed
remain specific if it is to be used at all.

Invasions will continue, and will continue to produce surprising and u
expected outcomes in a small proportion of cases. Deliberate releases of ar
sort should be discouraged. Proof that any particular release will be harmle
should be required of those wanting to do the release. It would be ove
optimistic to think that that will happen universally. Meanwhile, ecologis
need better studies of past invasions in order to give better advice on how
mitigate the effects of new ones.

References

Abernethy, K. (1994) The establishment of a hybrid zone between red and sika deer (genus *Cervus*). *Molecular Ecology*, **3**, 551–62.

Akeroyd, J.R. (1994) Some problems with introduced plants in the wild, in R.A. Perry and R.G. Ellis (eds) *The Common Ground of Wild and Cultivated Plants*, Botanical Society of the British Isles conference report number 22, National Museum of Wales, Cardiff, UK, pp. 31–40.

Allen, L.J.S., Allen, E.J., Kunst, C.R.G. and Sosebee, R.E. (1991) A diffusion model for dispersal of *Opuntia imbricata* (cholla) on rangeland. *Journal of Ecology*, **79**, 1123–35.

Altmann, M. (coordinator) (1993) Gene technology and biodiversity. *Experientia*, **49**, 187–234.

Amrine, J.W. Jr and Stasny, T.A. (1993) Biocontrol of multiflora rose, in B.N. McKnight (ed.) *Biological Pollution: The Control and Impact of Invasive Exotic Species*, Indiana Academy of Sciences, Indianapolis, IN, USA, pp. 9–22.

Anderson, A. (1990) *Prodigious Birds: Moas and Moa-hunting in Prehistoric New Zealand*, Cambridge University Press, Cambridge, UK.

Anderson, E. (1984) Who's who in the Pleistocene: a mammalian bestiary, in P.S. Martin and R.G. Klein (eds) *Quaternary Extinctions*, University of Arizona Press, Tucson, AZ, USA, pp. 40–89.

Anderson, R.M. (1994) The Croonian Lecture, 1994. Populations, infectious disease and immunity: a very nonlinear world. *Philosophical Transactions of the Royal Society B*, **346**, 457–505.

Anderson, R.M. and May, R.M. (1982) Coevolution of hosts and parasites. *Parasitology*, **85**, 411–29.

Anderson, R.M. and May, R.M. (1986) The invasion, persistence and spread of infectious diseases within plant and animal communities. *Philosophical Transactions of the Royal Society B*, **314**, 533–70.

Andow, D.A., Kareiva, P., Levin, S.A. and Okubo, A. (1990) Spread of invading organisms. *Landscape Ecology*, **4**, 177–88.

Andow, D.A., Kareiva, P.M., Levin, S.A. and Okubo, A. (1993) Spread of invading organisms: patterns of spread, in K.C. Kim and B.A. McPheron (eds) *Evolution of Insect Pests*, John Wiley & Sons, New York, USA, pp. 219–42.

Ankney, C.D., Dennis, D.G. and Bailey, R.C. (1987) Increasing mallards, decreasing American black ducks: coincidence or cause and effect? *Journal of Wildlife Management*, **51**, 523–9.

Anon. (1978) 'Mexican duck' not a Mexican duck. *Oryx*, **14**, 307.

Anon. (1994a) State Department of Agriculture leads the way to close door on harmful alien species. *The Nature Conservancy of Hawaii Newsletter*, Summer 1994, 7.

Anon. (1994b) Worldwide field trials of transgenic plants: 1986–1994. *The Gene Exchange*, **5**(3), 7.

Ashmole, N.P., Ashmole, M.J. and Simmons, K.E.L. (1994) Seabird conservation an feral cats on Ascension Island, South Atlantic, in D.N. Nettleship, J. Burger an M. Gochfeld (eds) *Seabirds on Islands, Threats, Case Studies and Action Plan* BirdLife International, Cambridge, UK, pp. 94–121.

Atkinson, I.A.E. (1985) The spread of commensal species of *Rattus* to oceanic island and their effects on island avifaunas, in P.J. Moors (ed.) *Conservation of Island Bird* International Council for Bird Preservation, Cambridge, UK, pp. 35–81.

Atkinson, T.C., Briffa, K.R. and Coope, G.R. (1987) Seasonal temperatures in Brita during the last 20,000 years, reconstructed using beetle remains. *Nature*, **325**, 587–9

Baker, A.J. and Moeed, A. (1987) Rapid genetic differentiation and founder effec in colonizing populations of common mynas *Acridotheres tristis*. *Evolution*, **4** 525–38.

Baker, H.G. (1965) Characters and modes of origin of weeds, in H.G. Baker ar G.L. Stebbins (eds) *The Genetics of Colonizing Species*, Academic Press, New Yor USA, pp. 147–72.

Baker, H.G. (1974) The evolution of weeds. *Annual Review of Ecology and Systematic* **5**, 1–24.

Baker, H.G. (1986) Patterns of plant invasions in North America, in H.A. Mooney ar J.A. Drake (eds) *Ecology of Biological Invasions of North America and Hawa* Ecological Studies 58, Springer-Verlag, New York, USA, pp. 44–57.

Banfield, A.W.F. (1974) *The Mammals of Canada*, University of Toronto Pres Toronto, Canada.

Bangerter, E.B. and Kent, D.H. (1957) *Veronica filiformis* Sm. in the British Isle *Proceedings of the Botanical Society of the British Isles*, **2**, 195–217.

Bangerter, E.B. and Kent, D.H. (1962) Further notes on Veronica filiformis. *Procee ings of the Botanical Society of the British Isles*, **4**, 384–97.

Bannerman, D.A. and Bannerman, W.M. (1965) *Birds of the Atlantic Islands, 2, History of the Birds of Madeira, the Desertas, and the Porto Santo Islands*, Oliver an Boyd, Edinburgh, UK.

Bannerman, D.A. and Bannerman, W.M. (1968) *Birds of the Atlantic Islands, 4, Histo of the Birds of the Cape Verde Islands*, Oliver and Boyd, Edinburgh, UK.

Barber, H.N. (1954) Genetic polymorphism of the rabbit in Tasmania. *Nature*, **17** 1227–9.

Barkai, A. and McQuaid, C. (1988) Predator-prey role reversal in a marine benth ecosystem. *Science*, **242**, 62–4.

Barrett, S.C.H. and Richardson, B.J. (1986) Genetic attributes of invading species, i R.H. Groves and J.J. Burdon (eds) *Ecology of Biological Invasions: an Australia Perspective*, Australian Academy of Science, Canberra, Australia, pp. 21–33.

Barry, J.P., Baxter, C.H., Sagarin, R.D. and Gilman, S.E. (1995) Climate-relate long-term faunal changes in a California rocky intertidal community. *Science*, **26** 672–5.

Bartlett, M.S. (1957) Measles periodicity and community size. *Journal of the Roy Statistical Society A*, **120**, 48–70.

Barton, N.H. and Charlesworth, B. (1984) Genetic revolutions, founder effects an speciation. *Annual Review of Ecology and Systematics*, **15**, 133–64.

Barton, N.H. and Hewitt, G.M. (1985) Analysis of hybrid zones. *Annual Review c Ecology and Systematics*, **16**, 113–48.

Beerling, D. and Perrins, J. (1993) *Impatiens glandulifera* Royle (*Impatiens royl* Walp.). Biological flora of the British Isles. *Journal of Ecology*, **81**, 367–82.

Beirne, B.P. (1975) Biological control attempts by introductions against pest insects in the field in Canada. *Canadian Entomologist*, **107**, 225–36.

Bennett, K.D. (1983) Postglacial population expansion of forest trees in Norfolk, UK. *Nature*, **303**, 164–7.

Bennett, W.A. (1990) Scale of investigation and the detection of competition: an example from the house sparrow and house finch introductions in North America. *American Naturalist*, **135**, 725–47.

Bertness, D. (1991) Zonation of *Spartina patens* and *Spartina alterniflora* in a New England salt marsh. *Ecology*, **72**, 138–48.

Birks, H.J.B. (1989) Holocene isochrone maps and patterns of tree-spreading in the British Isles. *Journal of Biogeography*, **16**, 503–40.

Birks, J.D.S. and Dunstone, N. (1991) Mink *Mustela vison*, in G.B. Corbet and S. Harris (eds) *The Handbook of British Mammals*, 3rd edn, Blackwell Scientific Publications, Oxford, UK, pp. 406–15.

Black, F.L. (1966) Measles endemicity in insular populations: critical community size and its implications. *Journal of Theoretical Biology*, **11**, 207–11.

Blackburn, T.M. and Lawton, J.H. (1994) Population abundance and body size in animal assemblages. *Philosophical Transactions of the Royal Society B*, **343**, 33–9.

Blakers, M., Davies, S.J.J.F. and Reilly, P.N. (1984) *The Atlas of Australian birds*, Melbourne University Press, Melbourne, Australia.

Blueweiss, L., Fox, H., Kudzma, V., Nakashima, D., Peters, R. and Sams, S. (1978) Relationships between body size and some life history parameters. *Oecologia*, **37**, 257–72.

Booth, E.S. (1968) *Mammals of Southern California*, University of California Press, Berkeley, California, USA.

Booth, T.H. (1985) A new method for assisting species selection. *Commonwealth Forestry Review*, **64**, 241–50.

Booth, T.H., Nix, H.A. and Hutchinson, M.F. (1987) Grid matching: a new method for homoclime analysis. *Agricultural and Forest Meteorology*, **39**, 241–55.

Boutin, S. and Birkenholz, D.E. (1987) *Wild Furbearer Management and Conservation in North America*, Ministry of Natural Resources (Ontario), Toronto, Canada.

Boyd Watt, H. (1923) On the American grey squirrel (*Sciurus carolinensis*) in the British Isles. *Essex Naturalist*, **20**, 189–205.

Bradshaw, A.D. (1991) The Croonian lecture, 1991. Genostasis and the limits to evolution. *Philosophical Transactions of the Royal Society B*, **333**, 289–305.

Braithwaite, R.W., Lonsdale, W.M. and Estbergs, J.A. (1989) Alien vegetation and native biota in tropical Australia: the impact of *Mimosa pigra*. *Biological Conservation*, **48**, 189–210.

Brill, W.J. (1985) Safety concerns and genetic engineering in agriculture. *Science*, **227**, 381–84.

Brown, A.H.D. and Marshall, D.R. (1981) Evolutionary changes accompanying colonization in plants, in G.G.E. Scudder and J.L. Reveal (eds) *Evolution Today*, Carnegie-Mellon University Press, Pittsburgh, PA, USA, pp. 351–63.

Brown, R.G.B. (1970) Fulmar distribution: a Canadian perspective. *Ibis*, **111**, 44–51.

Bundy, A. and Pitcher, T. (1995) An analysis of species changes in Lake Victoria: did the Nile perch act alone?, in T.J. Pitcher and P.J.B. Hart (eds) *The Impact of Species Changes in African Lakes*, Chapman & Hall, London, UK, pp. 111–135.

Burdon, J.J., Groves, R.H. and Cullen, J.M. (1981) The impact of biological control on the distribution and abundance of *Chondrilla juncea* in south-eastern Australia. *Journal of Applied Ecology*, **18**, 957–66.

Burt, W.H. and Grossenheider, R.P. (1976) *A Field Guide to the Mammals: North America North of Mexico*, 3rd edn, Houghton Miflin, Boston, MA, USA.

Caplan, A. and Van Montagu, M. (1990) Evolutionary consequences of modifying cultivated plants, in H.A. Mooney and G. Bernardi (eds) *Introduction of Genetically Modified Organisms into the Environment*, SCOPE 44, John Wiley & Sons, Chichester, UK, pp. 57–68.

Carlquist, S. (1965) *Island Life*, The Natural History Press, Garden City, NY, USA.

Carlquist, S. (1980) *Hawaii, a Natural History*, 2nd edn, Pacific Tropical Botanic Garden, Lawai, Kauai, HA, USA.

Carlton, J.T. (1987) Mechanisms and patterns of transoceanic marine biological invasions in the Pacific Ocean. *Bulletin of Marine Science*, **41**, 467–99.

Carlton, J.T. and Geller, J.B. (1993) Ecological roulette: the global transport of nonindigenous marine organisms. *Science*, **261**, 78–82.

Carroll, L. (1876) *The Hunting of the Snark*, Macmillan and Co., London, UK.

Carson, H.L. (1975) The genetics of speciation at the diploid level. *American Naturalist*, **109**, 73–92.

Case, T.J. (1990) Invasion resistance arises in strongly interacting species-rich model competitive systems. *Proceedings of the National Academy of Sciences (USA)*, **87**, 9610–14.

Cassels, R. (1984) Faunal extinction and prehistoric man in New Zealand and the Pacific islands, in P.S. Martin and R.G. Klein (eds) *Quaternary Extinctions*, University of Arizona Press, Tucson, AZ, USA, pp. 741–67.

Caswell, H. (1989) *Matrix Population Models*, Sinauer Associates, Sunderland, MA, USA.

Chambers, F.M. (ed.) (1993) *Climate Change and Human Impact on the Landscape*, Chapman & Hall, London, UK.

Chapman, N. (1991) Chinese muntjac *Muntiacus reevesi*, in G.B. Corbet and S. Harris (eds) *The Handbook of British Mammals*, 3rd edn, Blackwell Scientific Publications, Oxford, UK, pp. 526–32.

Chapman, N., Claydon, K., Claydon, M. and Harris, S. (1994) Muntjac in Britain: is there a need for a management strategy? *Deer*, **9**, 226–36.

Chapman, N., Harris, S. and Stanford, A. (1994) Reeve's muntjac *Muntiacus reevesi* in Britain: their history, spread, habitat selection, and the role of human intervention in accelerating their dispersal. *Mammal Review*, **24**, 113–60.

Charudattan, R. (1990) Biological control of aquatic weeds by means of fungi, in A.H. Pieterse and K.J. Murphy (eds) *Aquatic Weeds*, Oxford University Press, Oxford, UK, pp. 186–201.

Cheylan, G. (1991) Patterns of Pleistocene turnover, current distribution and speciation among Mediterranean mammals, in R.H. Groves and F. di Castri (eds) *Biogeography of Mediterranean Invasions*, Cambridge University Press, Cambridge, UK, pp. 227–62.

Clapham, A.R., Tutin, T.G. and Moore, D.M. (1987) *Flora of the British Isles*, 3rd edn, Cambridge University Press, Cambridge, UK.

Clark, D.A. (1984) Native land mammals, in R. Perry (ed.) *Key Environments. Galapagos*, International Union for Conservation of Nature, Geneva, Switzerland, pp. 225–31.

Cohen, J.E., Briand, F. and Newman, C.M. (1990) *Community Food Webs*, Biomathematics 20, Springer-Verlag, Berlin, Germany.

Colwell, R.K. (1992) Niche: a bifurcation in the conceptual lineage of the term i

E.F. Keller and E.A. Lloyd (eds) *Keywords in Evolutionary Biology*, Harvard University Press, Cambridge, Massachusetts, USA, pp. 241–8.

Combes, C. and Le Brun, L. (1990) Invasions by parasites in continental Europe, in F. di Castri, A.J. Hansen and M. Debussche (eds) *Biological Invasions in Europe and the Mediterranean Basin*, Kluwer Academic Publishers, Dordrecht, Netherlands, pp. 285–305.

Commonwealth of Australia (1984) *Biological Control Act of 1984*. No. 139 of 1984. Canberra, ACT, Australia.

Cook, C.D.K. (1990) Origin, autecology, and spread of some of the world's most troublesome aquatic weeds, in A.H. Pieterse and K.J. Murphy (eds) *Aquatic Weeds*, Oxford University Press, Oxford, UK, pp. 31–8.

Coope, G.R. (1986) The invasion and colonization of the North Atlantic islands: a palaeoecological solution to a biogeographic problem. *Philosophical Transactions of the Royal Society B*, **314**, 619–35.

Coope, G.R. (1987) The response of late quaternary insect communities to sudden climatic changes. *Symposia of the British Ecological Society*, **27**, 421–38.

Coope, G.R. (1990) The invasion of northern Europe during the Pleistocene by Mediterranean species of Coleoptera, in F. di Castri, A.J. Hansen and M. Debussche (eds) *Biological Invasions in Europe and the Mediterranean Basin*, Kluwer Academic Publishers, Dordrecht, Netherlands, pp. 203–15.

Cooper, A., Atkinson, I.A.E., Lee, W.G. and Worthy, T.H. (1993) Evolution of the moa and their effect on the New Zealand flora. *Trends in Ecology and Evolution*, **8**, 433–7.

Cooter, J. (ed.) (1991) *A Coleopterist's Handbook*, Amateur Entomologists' Society, Feltham, UK.

Corbet, G.B. and Clutton-Brock, J. (1984) Taxonomy and nomenclature, in I.L. Mason (ed.) *Evolution of Domesticated Animals*, Longman, London, UK, pp. 435–8.

Corbet, G.B. and Hill, J.E. (1991) *A World List of Mammalian Species*, 3rd edn, Natural History Museum Publications, London, UK.

Corbett, M. (1994) The ruddy duck dilemma. Part 2: in defence of the ruddy duck: a grass roots view. *Birdwatcher's Yearbook 1995*, 218–20.

Cornell, H.V. and Hawkins, B.A. (1993) Accumulation of native parasitoid species on introduced herbivores: a comparison of hosts as natives and hosts as invaders. *American Naturalist*, **141**, 847–65.

Courtney, W.R.Jr and Meffe, G.K. (1989) Small fishes in strange places: a review of introduced poecilids, in G.K Meffe and F.F. Snelson (eds) *Ecology and Evolution of Livebearing Fishes (Poecilidae)*, Prentice-Hall, Englewood Cliffs, NJ, USA, pp. 319–31.

Cowie, R.H. (1992) Evolution and extinction of Partulidae, endemic Pacific island snails. *Philosophical Transactions of the Royal Society B*, **335**, 167–91.

Cowling, R. (ed.) (1992) *The Ecology of Fynbos*, Oxford University Press, Cape Town, South Africa.

Craig, G.B. Jr (1993) The diaspora of the Asian tiger mosquito, in B.N. McKnight (ed.) *Biological Pollution: The Control and Impact of Invasive Exotic Species*, Indiana Academy of Science, Indianapolis, IN, USA, pp. 101–20.

Cramp, S. and Simmons, K.E.L. (eds) (1977) *Birds of the Western Palearctic*, Volume 1, Oxford University Press, Oxford, UK.

Cramp, S., Brooks, D.J., Dunn, E., Gillmor, R., Hollom, P.A.D., Hudson, R., Nicholson, E.M., Ogilvie, M.A., Olney, P.T.S., Roselaar, C.S., Simmons, K.E.G.,

Voous, K.H., Wallace, D.M., Wattel, J. and Wilson, M.G. (eds) (1985) *Birds of the Western Palearctic*, Volume 4, Oxford University Press, Oxford, UK.

Crawley, M.J. (1986) The population biology of invaders. *Philosophical Transactions of the Royal Society B*, **314**, 711–31.

Crawley, M.J. (1987) What makes a community invasible? *Symposia of the British Ecological Society*, **26**, 429–53.

Crawley, M.J. (1989) Chance and timing in biological invasions, in J.A. Drake, H.A. Mooney, F. di Castri, R.H. Groves, F.J. Kruger, M. Rejmánek and M. Williamson (eds) *Biological Invasions, a Global Perspective*, John Wiley & Sons, Chichester, UK, pp. 407–23.

Crichton, M. (1974) *Provisional Distribution Maps of Amphibians, Reptiles and Mammals in Ireland.* Folens, Dublin, Ireland.

Crisp, D.J. (1958) The spread of *Elminius modestus* Darwin in north-west Europe. *Journal of the Marine Biological Association*, **37**, 483–520.

Crompton, C.W., McNeill, J., Stahevitch, A.E. and Wotjas, W.A. (1988) *Preliminary Inventory of Canadian weeds*, Technical Bulletin 1988-9E, Research Branch, Agriculture Canada, Ottawa, Canada.

Cronk, Q.C.B. and Fuller, J.L. (1995) *Plant Invaders*, Chapman & Hall, London, UK.

Cullen, J.M. and Delfosse, E.S. (1991) Progress and prospects in biological control of weeds. *Proceedings of the 9th Australian Weeds Conference*, 452–62.

Culotta, E. (1994) The weeds that swallowed the west. *Science*, **265**, 1178–9.

Daehler, C.C. and Strong, D.R. Status, prediction, and prevention of introduced cordgrasses (*Spartina* spp.) in Pacific estuaries, USA. *Biological Conservation* (in press).

Daehler, C.C. and Strong, D.R.Jr (1993) Prediction and biological invasions. *Trends in Ecology and Evolution*, **8**, 380.

Damuth, J. (1991) Of size and abundance. *Nature*, **351**, 268–9.

Dandy, J.E. (1969) *Watsonian Vice-counties of Great Britain*, The Ray Society, London, UK.

Daniel, M. and Baker A. (1986) *Collins Guide to the Mammals of New Zealand*, Collins, Auckland, New Zealand.

Dansie, O. (1977) Muntjac *Muntiacus reevesi*, in G.B. Corbet and H.N. Southern (eds) *The Handbook of British mammals*, 2nd edn, Blackwell Scientific Publications, Oxford, UK, pp. 447–51.

Dansie, O. (1983) *Muntjac*, British Deer Society publication no. 2, British Deer Society, Warminster, UK.

Darwin, C. (1860) *On the Origin of Species by Means of Natural Selection*, 2nd edn, John Murray, London, UK.

Davis, M.B. (1983) Quaternary history of deciduous forests of eastern North America and Europe. *Annals of the Missouri Botanic Garden*, **70**, 550–63.

Davis, S.J.M. (1987) *The Archaeology of Animals*, B.T. Batsford, London, UK.

Dawson, J.C. (1984) *A Statistical Analysis of Species Characteristics Affecting the Success of Bird Introductions*. BSc. Thesis, University of York.

DeBach, P. (1965) Some biological and ecological phenomena associated with colonizing entomophagous insects, in H.G. Baker and G.L. Stebbins (eds) *The Genetics of Colonizing Species*, Academic Press, New York, USA, pp. 287–306.

Delcourt, P.A. and Delcourt, H.R. (1987) *Long-term Forest Dynamics of the Temperate Zone*, Ecological Studies 63, Springer-Verlag, New York, USA.

Delfosse, E.S. and Cullen, J.M (1985) CSIRO Division of Entomology submission t the inquiries into biological control of *Echium plantagineum* L., Paterson's curse/salvation Jane. *Plant Protection Quarterly*, **1**, 24–40.

Dennill, G.G., Donnelly, D. and Chown, S.L (1993) Expansion of host range of a biological control agent *Trichilogaster acaciaelongifoliae* (Pteromalidae) released against the weed *Acacia longifolia* in South Africa. *Agriculture, Ecosystems and Environment*, **43**, 1–10.

di Castri, F. (1989) History of biological invasions with special emphasis on the old world, in J.A. Drake, H.A. Mooney, F. di Castri, R.H. Groves, F.J. Kruger, M. Rejmánek and M. Williamson (eds) *Biological Invasions, a Global Perspective*, John Wiley & Sons, Chichester, UK, pp. 1–30.

Diamond, J.M. (1984) Historic extinction: a Rosetta stone for understanding prehistoric extinctions, in P.S. Martin and R.G. Klein (eds) *Quaternary Extinctions*, University of Arizona Press, Tucson, AZ, USA, pp. 824–62.

Dobson, A.P. and May, R.M. (1986) Patterns of invasion by pathogens and parasites, in H.A. Mooney and J.A. Drake (eds) *Ecology of Biological Invasions of North America and Hawaii*, Ecological Studies 58, Springer- Verlag, New York, USA, pp. 58–76.

Dodds, D. (1983) Terrestrial mammals, in G.R. South (ed.) *Biogeography and Ecology of the Island of Newfoundland*, Monographiae Biologicae 48, Dr W. Junk, The Hague, Netherlands, pp. 509–50.

Donne, J. (1624) *Devotions upon Emergent Occassions*, London, England.

Dovers, S.R. and Handmer, J.W. (1995) Ignorance, the precautionary principle, and sustainability. *Ambio*, **24**, 92–7.

Downes, J.A. (1988) The post-glacial colonization of the North Atlantic islands. *Memoirs of the Entomological Society of Canada*, **144**, 55–92.

Drake, J.A. (1990) The mechanics of community assembly and succession. *Journal of Theoretical Biology*, **147**, 213–33.

Drake, J.A., Mooney, H.A., di Castri, F., Groves, R.H., Kruger, F.J., Rejmánek, M. and Williamson, M. (eds) (1989) *Biological Invasions, a Global Perspective,* John Wiley & Sons, Chichester, UK.

Drodjevic, M.A., Schofield, P.R. and Rolfe, B.G. (1985) Tn5 mutagenesis of *Rhizobium trifolii* host-specific nodulation genes result in mutants with altered host-range ability. *Molecular and General Genetics*, **200**, 463–71.

Dybbro, T. (1976) *De Danske Ynglefugles Udbredelse*. Dansk Ornithologisk Forening, København, Denmark.

Ehrlich, P.R. (1986) Which animal will invade?, in H.A. Mooney and J.A. Drake (eds) *Ecology of Biological Invasions of North America and Hawaii.* Ecological Studies 58, Springer-Verlag, New York, USA, pp. 79–95.

Ehrlich, P.R. (1989) Attributes of invaders and the invading process: vertebrates, in J.A. Drake, H.A. Mooney, F. di Castri, R.H. Groves, F.J. Kruger, M. Rejmánek, and M. Williamson (eds) *Biological Invasions, a Global Perspective*, SCOPE 37, John Wiley & Sons, Chichester, UK, pp. 315–28.

Elias, S.A. (1994) *Quaternary Insects and Their Environments*, Smithsonian Institution Press, Washington, DC, USA.

Elton, C. (1927) *Animal Ecology*, Sidgwick and Jackson, London, UK.

Elton, C.S. (1958) *The Ecology of Invasions by Animals and Plants,* Methuen, London, UK.

Emmet, A.M. and Heath, J. (eds) (1989) *The Moths and Butterflies of Great Britain and Ireland,* 7(1) *Hesperiidae-Nymphalidae, the Butterflies,* Harley Books, Colchester, UK.

Engbring, J. and Fritts, T.H. (1988) Demise of an insular avifauna: the brown t snake on Guam. *Transactions of the Western Section of the Wildlife Society*, 31–7.

Evans, H.F. and Fielding, N.J. (1994) Integrated management of *Dendroctonus mic* in the UK. *Forest Ecology and Management*, **65**, 17–30.

Evans, P.G.H. (1985a) The seabirds of Greenland: their status and conservation, J.P. Croxall, P.G.H. Evans and R.W. Schreiber (eds) *Status and Conservation of World's Seabirds*, Technical Publication 2, International Council for Bird Preser tion, Cambridge, UK, pp. 49–84.

Evans, P.G.H. (1985b) Status and conservation of seabirds in northwestern Europe J.P Croxall, P.G.H. Evans and R.W. Schreiber (eds) *Status and Conservation of World's seabirds*, Technical Publication 2, International Council for Bird Preser tion, Cambridge, UK, pp. 293–322.

Ewel, J.J. (1986) Invasibility: lessons from south Florida, in H.A. Mooney a J.A. Drake (eds) *Ecology of Biological Invasions of North America and Haw* Ecological Studies 58, Springer-Verlag, New York, USA, pp. 214–30.

Falla, R.A., Sibson, R.B and Turbott, E.G. (1978) *Collins Guide to the Birds of N Zealand,* Collins, Auckland, NZ.

Fauquet, C.M. (1994) Taxonomy and classification – general, in R.G. Webster a A. Granoff (eds) *Encyclopedia of Virology*, Academic Press, London, U pp. 1396–410.

Fenchel, T. (1974) Intrinsic rate of natural increase: the relationship with body si *Oecologia*, **14**, 317–26.

Fenner, F. (1965) Myxoma virus and *Oryctolagus cuniculus*: two colonizing species, H.G. Baker and G.L. Stebbins (eds) *The Genetics of Colonizing Species*, Acader Press, New York, USA, pp. 485–501.

Fenner, F. (1983) Biological control as exemplified by smallpox eradication a myxomatosis. The Florey Lecture, 1983. *Proceedings of the Royal Society B*, **2** 259–85.

Fenner, F. and Ratcliffe, F.N. (1965) *Myxomatosis*, Cambridge University Pre Cambridge, UK.

Fenner, F. and Ross, J (1994) Myxomatosis, in H.V. Thompson and C.M King (e *The European Rabbit, the History and Biology of a Successful Colonizer*, Oxfc University Press, Oxford, UK, pp. 205–39.

Fisher, J. (1952) *The Fulmar*, New Naturalist Monograph 6, Collins, London, UK.

Fisher, J. (1966) The fulmar population of Britain and Ireland, 1959. *Bird Study*, 5–76.

Fisher, J. and Lockley, R.M. (1954) *Sea-birds*, New Naturalist 28, Collins, Lond UK.

Fisher, R.A. (1937) The wave of advance of advantageous genes. *Annals of Eugen* **7**, 355–69.

Fitter, A. (1978) *An Atlas of the Wild Flowers of Britain and Northern Europe*, Colli London, UK.

Fitter, A.H. and Peat, H.J. (1994) The Ecological Flora Database. *Journal of Ecolo* **82**, 415–25.

Fitter, R.S.R. (1958) *The Ark in our Midst*, Collins, London, UK.

Flux, J.E.C. (1994) World distribution, in H.V. Thompson and C.M King (eds) *T European Rabbit, the History and Biology of a Successful Colonizer*, Oxford Unive ity Press, Oxford, UK, pp. 8–21.

Flux, J.E.C. and Fullagar, P.J. (1992) World distribution of the rabbit *Oryctolagus cuniculus* on islands. *Mammal Review*, **22**, 151–205.

Fornell, T.C. (1990) Widespread adventive plants of Catalonia, in F. di Castri, A.J. Hansen and M. Debussche (eds) *Biological Invasions in Europe and the Mediterranean Basin*, Kluwer Academic Publishers, Dordrecht, Netherlands, pp. 85–104.

Fritts, T.H. (1988) *The Brown Tree Snake,* Boiga irregularis*, A Threat to Pacific Islands.* US Department of the Interior Fish and Wildlife Service Biological Report 88 (31), Washington, DC, USA.

Fritts, T.H., McCoid, M.J. and Haddock, R.L. (1994) Symptoms and circumstances associated with bites by the brown tree snake (Colubridae: *Boiga irregularis*) on Guam. *Journal of Herpetology*, **28**, 27–33.

Funasaki, G.Y., Lai, P.-Y., Nakahara, L.M., Beardsley, J.W. and Ota, A.K. (1988) A review of biological control introductions in Hawaii: 1890 to 1985. *Proceedings, Hawaiian Entomological Society*, **28**, 105–60.

Furness, R.W. and Todd, C.M. (1984) Diets and feeding ecology of fulmars *Fulmarus glacialis* during the breeding season: a comparison between St Kilda and Shetland colonies. *Ibis*, **126**, 379–87.

Galil, B.S. (1993) Lessepsian migration: new findings on the foremost anthropogenic change in the Levant basin fauna, in N.F.R. Della Croce (ed.) *Symposium Mediterranean Seas 2000*, Istituto Scienze Ambientali Marine, Università di Genova, Santa Margherita Ligure, Italy, pp. 307–23.

Garnery, L., Cornuet, J.-M. and Solignac, M. (1992) Evolutionary history of the honey bee *Apis mellifera* inferred from mitochondrial DNA analysis. *Molecular Ecology*, **1**, 145–54.

Garvey, T. (1935) The muskrat in Saorstat Eireann. *Journal of the Department of Agriculture, Dublin*, **35**, 189–95.

Gaston, K.J. (1988) The intrinsic rates of increase of insects of different sizes. *Ecological Entomology*, **14**, 399–409.

Gaston, K.J. (1994) *Rarity*, Chapman & Hall, London, UK.

Gibb, J.A. and Williams, J.M. (1994) The rabbit in New Zealand, in H.V. Thompson and C.M King (eds) *The European Rabbit, the History and Biology of a Successful Colonizer*, Oxford University Press, Oxford, UK, pp. 158–204.

Gibbons, D.W., Reid, J.B. and Chapman, R.A. (1993) *The New Atlas of Breeding Birds in Britain and Ireland: 1988–1991*, Poyser, London, UK.

Gibbs, J.N. and Wainhouse, D. (1986) Spread of forest pests and pathogens in the northern hemisphere. *Forestry*, **59**, 141–53.

Gillespie, G.D. (1985) Hybridization, introgression, and morphometric differentiation between mallard (*Anas platyrhynchos*) and grey duck (*Anas superciliosa*) in Otago, New Zealand. *Auk*, **102**, 459–69.

Gilpin, M. (1990) Ecological prediction. *Science*, **248**, 88–9.

Gledhill, T., Sutcliffe, D.W. and Williams, W.D. (1976) *Key to British Freshwater Crustacea: Malacostraca*, Scientific Publication 32, Freshwater Biological Association, Ambleside, UK.

Glutz von Blotzheim, U.N. (ed.) (1980) *Handbuch der Vögel Mitteleuropas*, Band 9, Columbiformes – Piciformes, Akademische Verlagsgesellschaft, Wiesbaden, Germany.

Godfrey, W.E. (1966) *The Birds of Canada*, National Museums of Canada, Ottawa, Canada.

Goeden, R.D. (1983) Critique and revision of Harris' scoring system for selection insect agents in biological control of weeds. *Protection Ecology*, **5**, 287–301.

Goeden, R.D., Ricker, D.W and Hawkins, B.A. (1986) Ethological and genetic diff ences among three biotypes of *Rhinocyllus conicus* (Coleoptera: Curculionid introduced into North America for the biological control of asteraceous thist *Proceedings of the VI International Symposium on the Biological Control of Wee* 181–9.

Golani, D. (1993) The sandy shore of the Red Sea – launching pad for Lessepsian (S Canal) migrant fish colonizers of the eastern Mediterranean. *Journal of Biog graphy*, **20**, 579–85.

Gonçalves, L.S., Stort, A.C. and De Jong, D. (1991) Beekeeping in Brazil, in M. Spiv D.J.C. Fletcher and M.D. Breed (eds) *The 'African' honey bee*, Westview Pr Boulder, CO, USA, pp. 359–72.

Goodman, P.J., Braybooks, E.M., Lambert, J.M. and Marchant, C.J. (1969) Biol ical flora of the British Isles: *Spartina* Screb. *Journal of Ecology*, **57**, 285–313.

Goodwin, D. (1970) *Pigeons and Doves of the World*, British Museum (Natu History), London, UK.

Gosling, L.M. and Baker, S.J.(1991) Coypu *Myocaster coypus*, in G.B. Corbet a S. Harris (eds) *The Handbook of British Mammals*, 3rd edn, Blackwell Scient Publications, Oxford, UK, pp. 267–75.

Gottlieb, L.D. (1984) Genetics and morphological evolution in plants. *American N uralist*, **123**, 681–709.

Gray, A.J. (1986) Do invading species have definable characteristics? *Philosoph Transactions of the Royal Society B*, **314**, 655–74 (and 654).

Gray, A.J., Benham, P.E.M. and Raybould, A.F. (1990) *Spartina anglica* – the evo tionary and ecological background, in A.J. Gray and P.E.M. Benham (eds) Spart anglica – *a research review*, ITE research publication no. 2, HMSO, London, U pp. 5–10.

Gray, A.J., Marshall, D.F. and Raybould, A.F. (1991) A century of evolution *Spartina anglica*. *Advances in Ecological Research*, **21**, 1–62.

Gregory, R.D. and Blackburn, T.M. (1995) Abundance and body size in British bi reconciling regional and ecological densities. *Oikos*, **72**, 151–4.

Griesmer, J.R. (1992) Niche: historical perspectives, in E.F. Keller and E.A. Lloyd (e *Keywords in Evolutionary Biology*, Harvard University Press, Cambridge, Mas chusetts, USA, pp. 231–40.

Griffin, J.R. and Critchfield, W.B. (1976) *The Distribution of Forest Trees in Califor* USDA Forest Service Research paper PSW-82, reprint with supplement, Berkel California, USA.

Griffith, B., Scott, M.J., Carpenter, J.W. and Reed, C. (1989) Translocation as a spe conservation tool: status and strategy. *Science*, **245**, 477–80.

Grime, J.P. (1987) Dominant and subordinate components of plant communit implications for succession, stability and diversity. *Symposia of the British Ecolog Society*, **26**, 413–28.

Grimm, V., Schmidt, E. and Wissel, C. (1992) On the application of stability conce in ecology. *Ecological Modelling*, **63**, 143–61.

Grinnell, J. (1917) The niche-relationships of the California thrasher. *Auk*, **34**, 427–

Groombridge, B. (ed.) (1992) *Global Biodiversity*, Chapman & Hall, London, UK.

Grosholz, E.D. Contrasting rates of spread for introduced species in terrestrial a marine species. *Ecology* (in press).

Groves, R.H. (1989) Ecological control of invasive terrestrial plants, in J.A. Drake, H.A. Mooney, F. di Castri, R.H. Groves, F.J. Kruger, M. Rejmánek and M. Williamson (eds) *Biological Invasions, a Global Perspective*, SCOPE 37, John Wiley & Sons, Chichester, UK, pp. 437–61.

Grubb, P.J. (1987) Some generalizing ideas about colonization and succession in green plants and fungi. *Symposia of the British Ecological Society*, **26**, 81–102.

Guilday, J.E. (1984) Pleistocene extinction and environmental change: case study of the Appalachians, in P.S. Martin and R.G. Klein (eds) *Quaternary Extinctions*, University of Arizona Press, Tucson, AZ, USA, pp. 259–98.

Gupta, R.K. (1989) *The Living Himalayas, Volume 2, Aspects of Plant Explorations and Phytogeography*, Today & Tomorrow's Printers and Publishers, New Delhi, India.

Gurnell, J. (1991) Genus *Sciurus*, in G.B. Corbet and S. Harris (eds) *The Handbook of British Mammals*, 3rd edn, Blackwell Scientific Publications, Oxford, UK, pp. 177–91.

Hall, E.R. (1981) *The Mammals of North America*, 2nd edn, John Wiley & Sons, New York, NY, USA.

Hall, H.G. (1991) Genetic characterization of honey bees through DNA analysis, in M. Spivak, D.J.C. Fletcher and M.D. Breed (eds) *The 'African' Honey Bee*, Westview Press, Boulder, CO, USA, pp. 45–76.

Hall, S.J and Raffaelli, D.G. (1993) Food webs: theory and reality. *Advances in Ecological Research*, **24**, 187–239.

Hannah, L., Lohse, D., Hutchinson, C., Carr, J.L. and Lankerani, A. (1994) A preliminary inventory of human disturbance of world ecosystems. *Ambio*, **23**, 246–50.

Hanski, I. and Camberfort, Y. (eds) (1991) *Dung beetle ecology*, Princeton University Press, Princeton, NJ, USA.

Hanski, I., Kouki, J. and Halkka, A. (1993) Three explanations of the positive relationship between distribution and abundance of species, in R.E. Ricklefs and D. Schluter (eds) *Species Diversity in Ecological Communities*, The University of Chicago Press, Chicago, IL, USA, pp. 108–16.

Harland, W.B., Armstrong, R.L., Cox, A.V., Craig, L.E., Smith, A.G. and Smith, D.G. (1990) *A Geologic Time Scale, 1989*, Cambridge University Press, Cambridge, UK.

Harley, K.L.S. and Forno, I.W. (1990) Biological control of aquatic weeds by means of arthropods, in A.H. Pieterse and K.J. Murphy (eds) *Aquatic Weeds*, Oxford University Press, Oxford, UK, pp. 177–86.

Harris, P. (1985) Biocontrol of weeds: bureaucrats, botanists, beekeepers and other bottlenecks, in E.S. Delfosse (ed.) *Proceedings of the VI International Symposium on the Biological Control of Weeds*, Agriculture Canada, Ottawa, Canada, pp. 3–12.

Harrison, C. (1982) *An Atlas of the Birds of the Western Palaearctic*, Collins, London, UK.

Harrison, J.F. and Hall, H.G. (1993) African-European honeybee hybrids have low non-intermediate metabolic capacities. *Nature*, **363**, 258–60.

Hastings, A. Models of spatial spread: a synthesis. *Biological Conservation* (in press).

Hawaii Audubon Society (1993) *Hawaii's Birds,* 4th edn, Hawaii Audubon Society, Honolulu, HA, USA.

Hawkes, N. (1993) The culling fields. *The Times Magazine* for October 16, 50.

Hawkins, B.A. and Gross, P. (1992) Species richness and population limitation in insect parasitoid-host systems. *American Naturalist*, **139**, 417–23.

Hengeveld, R. (1989) *Dynamics of Biological Invasions*, Chapman & Hall, London, UK.

Herbold, B. and Moyle, P.B. (1986) Introduced species and vacant niches. *American Naturalist*, **128**, 751–60.

Heywood, V.H. (1989) Patterns, extents and modes of invasions by terrestrial plants, in J.A. Drake, H.A. Mooney, F. di Castri, R.H. Groves, F.J. Kruger, M. Rejmánek and M. Williamson (eds) *Biological Invasions, a Global Perspective*, John Wiley & Sons, Chichester, UK, pp. 31–60.

Hickley, P. (1986) Invasion by zander and the management of fish stocks. *Philosophical Transactions of the Royal Society B*, **314**, 571–82.

Hindar, K., Ryman, N. and Utter, F. (1991) Genetic effects of cultured fish on natural fish populations. *Canadian Journal of Fisheries and Aquatic Sciences*, **48**, 945–57.

Hirsch, M. and Smale, S. (1974) *Differential Equations, Dynamical Systems and Linear Algebra*, Academic Press, New York, NY, USA.

Hobbs, J.N. (1961) The birds of south-west New South Wales. *Emu*, **61**, 21–55.

Hoeck, H.N. (1984) Introduced fauna, in R. Perry (ed.) *Galapagos*, Pergamon Press, Oxford, UK, pp. 233–45.

Hofbauer, J. and Sigmund, K. (1988) *The Theory of Evolution and Dynamical Systems*, London Mathematical Society Student Texts 7, Cambridge University Press, Cambridge, UK.

Hoffman, M. (1958) *Die Bisamratte*, Academische Verlagsgesellschaft, Leipzig, Germany.

Holdgate, M.W. (1986) Summary and conclusions: characteristics and consequences of biological invasions. *Philosophical Transactions of the Royal Society B*, **314**, 733–42.

Holdich, H. (1991) The native crayfish and threats to its existence. *British Wildlife*, **2**, 141–51.

Holmes, E.E. (1993) Are diffusion models too simple? A comparison with telegraph models of invasion. *American Naturalist*, **142**, 779–95.

Holmes, E.E., Lewis, M.A. Banks, J.E. and Veit, R.R. (1994) Partial differential equations in ecology: spatial interactions and population dynamics. *Ecology*, **75**, 17–29.

Holmes, J. (1994) The ruddy duck dilemma. Part 1: the UK ruddy duck working group. *The Birdwatcher's Yearbook 1995*, 215–18.

Hopper, K.R., Roush, R.T. and Powell, W. (1993) Management of genetics of biological control introductions. *Annual Review of Entomology*, **38**, 27–51.

Howard, R and Moore, A. (1991) *A Complete Checklist of the Birds of the World*, 2nd edn, Academic Press, London, UK.

Howarth, F.G. and Medeiros, A.C. (1989) Non-native invertebrates, in C.P Stone and D.B. Stone (eds), *Conservation Biology in Hawai'i*, University of Hawaii Cooperative National Park Resources Study Unit, Honolulu, HA, USA, pp. 82–7.

Hughes, L., Dunlop, M., French, K., Leishman, M.R., Rice, B., Rodgerson, L. and Westoby, M. (1994) Predicting dispersal spectra: a minimal set of hypotheses based on plant attributes. *Journal of Ecology*, **82**, 933–50.

Hultén, E. (1971) *Atlas över Växternas Utbredning i Norden*, Generalstaben Litografiska Anstalts Förlag, Stockholm, Sweden.

Hultén, E. and Fries, M. (1986) *Atlas of North European Vascular Plants*, Koeltz Scientific Books, Königstein, Germany.

Huntley, B. (1993) Rapid early-Holocene migration and high abundance of hazel (*Corylus avellana* L.): alternative hypotheses, in F.M. Chambers (ed.), *Climate Change and Human Impact on the Landscape*, Chapman & Hall, London, UK, pp. 205–15.

Huntley, B. and Birks, H.J.B. (1983) *An Atlas of Past and Present Pollen Maps for Europe: 0–13000 Years Ago*, Cambridge University Press, Cambridge, UK.

Hutchinson, G.E. (1957) Concluding remarks. *Cold Spring Harbor Symposia on Quantitative Biology*, **22**, 415–27.

Hutson, V. and Law, R. (1985) Permanent coexistence in general models of three interacting species. *Journal of Mathematical Biology*, **21**, 285–98.

Hutson, V. and Schmitt, K. (1992) Permanence and the dynamics of biological systems. *Mathematical Biosciences*, **111**, 1–71.

Huyer, A. (1983) Coastal upwelling in the California current system. *Progress in Oceanography*, **12**, 259–84.

Isberg, R.R. and Falkow, S. (1985) A single genetic locus encoded by *Yersinia pseudotuberculosis* permits invasion of cultured animal cells of *Escherichia coli* K-12. *Nature*, **317**, 262–4.

Jaksic, F.M. and Fuentes, E.R. (1991) Ecology of a successful invader: the European rabbit in central Chile, in R.H. Groves and F. di Castri (eds) *Biogeography of Mediterranean Invasions*, Cambridge University Press, Cambridge, UK, pp. 273–84.

James, H.F. and Olson, S.L. (1991) Descriptions of thirty-two new species of birds from the Hawaiian Islands: Part II. Passeriformes. *Ornithological Monographs of the American Ornithologists' Union*, **46**, 1–88.

Johnson, L. and Padilla, D.K. Geographic spread of exotic species: ecological lessons and opportunities from the invasion of zebra mussel *Dreissena polymorpha*. *Biological Conservation* (in press).

Johnson, M.S., Murray, J. and Clarke, B. (1993) The ecological genetics and adaptive radiation of *Partula* on Moorea. *Oxford Surveys in Evolutionary Biology*, **9**, 167–238.

Johnson, W.C. and Webb, T. III (1989) The role of blue jays (*Cyanocitta cristata* L.) in the postglacial dispersal of fagaceous trees in eastern North America. *Journal of Biogeography*, **16**, 561–71.

Johnstone, J., Scott, A. and Chadwick, H.C. (1924) *The Marine Plankton*, Liverpool University Press, Liverpool, UK.

Johnstone, R.F. and Klitz, L.J. (1977) Variation and evolution in a granivorous bird; the house sparrow, in J. Pinkowski and S.C. Kendeigh (eds), *Granivorous Birds in Ecosystems*, Cambridge University Press, Cambridge, UK, pp. 15–51.

Juvik, J.O. and Juvik, S.P. (1992) Mullein (*Verbascum thapsus*): the spread and adaptation of a temperate weed in the montane tropics, in C.P. Stone, C.W. Smith and J.T. Tunison (eds) *Alien Plant Invasions in Native Ecosystems of Hawai'i: Management and Research*, University of Hawaii Cooperative National Park Resources Study Unit, Hononlulu, HA, USA, pp. 254–70.

Kear, J. (1990) *Man and Wildfowl*, T. & A.D. Poyser, London, UK.

Kenward, R.E. and Holm, J.L. (1993) On the replacement of the red squirrel in Britain: a phytotoxic explanation. *Proceedings of the Royal Society B*, **251**, 187–94.

Kideys, A.E. (1994) Recent dramatic changes in the Black sea ecosytem: the reason for the sharp decline in Turkish anchovy fisheries. *Journal of Marine Systems*, **5**, 171–81.

Kilbourne, E.D. (1980) Influenza: viral determinants of the pathogenicity and epidemicity of an invariant disease of variable occurrence. *Philosophical Transactions of the Royal Society B*, **288**, 291–7.

Kilbourne, E.D. (1987) *Influenza*, Plenum, New York, NY, USA.

King, C. (1984) *Immigrant Killers*, Oxford University Press, Auckland, NZ.

King, W.B. (1985) Island birds: will the future repeat the past?, in P.J. Moors (ed.)

Conservation of Island Birds, Technical Publication No. 3, International Council f Bird Preservation, Cambridge, UK, pp. 3–15.

Knowlton, N., Weigt, L.A., Solórzano, L.A., Mills, D.K. and Bermingham, E. (19 Divergence in proteins, mitochondrial DNA, and reproductive compatibility acr the Isthmus of Panama. *Science*, **260**, 1629–32.

Kolb, H.H. (1994) Rabbit *Oryctolagus cuniculus* populations in Scotland since t introduction of myxomatosis. *Mammal Review*, **24**, 41–8.

Kornaś, J. (1990) Plant invasions of central Europe: historical and ecological aspec in F. di Castri, A.J. Hansen and M. Debussche (eds) *Biological Invasions in Eur and the Mediterranean Basin*. Kluwer Academic Publishers, Dordrecht, Netherlan pp. 19–36.

Kowarik, I. (1995) Time lags in biological invasions with regard to the success a failure of alien species, in P. Pyšek, K. Prach, M. Rejmánek and M. Wade (eds) *Pl Invasions – General aspects and Special Problems*. SPB Academic Publishi Amsterdam, Netherlands, pp. 15–38.

Krattiger A.F. and Rosemarin A.(eds) (1994) *Biosafety for Sustainable Agricultu Sharing Biotechnology Regulatory Experiences of the Western Hemisphere*, ISA Ithaca, NY, USA and SEI, Stockholm, Sweden.

Krattiger, A.F. (1994) The field testing and commercialization of genetically enginee plants: a review of worldwide data (1986 to 1993/94), in A. Krattiger a A. Rosemarin (eds) *Biosafety for Sustainable Agriculture: Sharing Biotechnol Regulatory Experiences of the Western Hemisphere*, ISAA, Ithaca, NY, USA a SEI, Stockholm, Sweden, pp. 247–66.

Lack, P. (1986) *The Atlas of Wintering Birds in Britain and Ireland*, T. & A.D. Poys Calton, UK.

Lack, P. (1992) *Birds on Lowland Farms*, HMSO, London, UK.

Law, R. and Blackford, J.C. (1992) Self-assembling food webs: a global view coexistence of species in Lotka-Volterra communities. *Ecology*, **73**, 567–78.

Law, R. and Morton, R.D. (1993) Alternative permanent states of ecological comm nities. *Ecology*, **74**, 1347–61.

Law, R. and Morton, R.D. Permanence and the assembly of ecological communiti *Ecology* (in press).

Law, R. and Watkinson, A.R. (1989) Competition. *Symposia of the British Ecologi Society*, **29**, 243–84.

Lawton, J.H. (1990) Biological control of plants: a review of generalisations, rules, a principles using insects as agents, in C. Bassett, L.J. Whitehouse and J.A. Zabkiew J.A. (eds) *Alternatives to the Chemical Control of Weeds*, New Zealand Ministry Forestry, FRI Bulletin 155, Wellington, New Zealand, pp. 3–17.

Lawton, J.H. (1992) Feeble links in food webs. *Nature*, **355**, 19–20.

Lawton, J.H. (1993) Range, population abundance and conservation. *Trends in Ecolo and Evolution*, **8**, 409–13.

Lawton, J.H. and Brown, K.C. (1986) The population and community ecology invading insects. *Philosophical Transactions of the Royal Society B*, **314**, 606–17.

Leader-Williams, N. (1988) *Reindeer on South Georgia*, Cambridge University Pre Cambridge, UK.

Lever, C. (1994) *Naturalized Animals*, T. & A.D. Poyser, London, UK.

Levin, S.A. (1988) Sea otters and nearshore benthic communities: a theoretical perspe ive, in G.R. VanBlaricom and J.A. Estes, *The community ecology of sea otte* Ecological Studies 65, Springer-Verlag, New York, USA, pp. 202–9.

Levings, C.S. III (1990) The Texas cytoplasm of maize: cytoplasmic male sterility and disease susceptibility. *Science*, **250**, 942–7.

Lewis, M.A. and Kareiva, P. (1993) Allee dynamics and the spread of invading organisms. *Theoretical Population Biology*, **43**, 141–58.

Lister, A.M. (1989) Rapid dwarfing of red deer on Jersey in the Last Interglacial. *Nature*, **342**, 539–42.

Litvinenko, N and Shibaev, Y. (1991) Status and conservation of seabirds nesting in southeast U.S.S.R., in J.P. Croxall (ed.) *Seabird Status and Conservation: a Supplement*, Technical Publication 11, International Council for Bird Preservation, Cambridge, UK, pp. 175–204.

Lloyd, C., Tasker, M.L. and Partridge, K. (1991) *The Status of Seabirds in Britain and Ireland*, Poyser, London, UK.

Lloyd, H.G. (1983) Past and present distribution of red and grey squirrels. *Mammal Review*, **13**, 69–80.

Lobo, J.A. (1995) Morphometric, isozymic and mitochondrial variability of Africanized honeybees in Costa Rica. *Heredity*, **75**, 133–41.

Lockley, R.M. (1947) *Letters from Skokholm*, Dent, London, UK.

Lohmeyer, W. and Sukopp, H. (1992) Agriophyten in der Vegetation Mitteleuropas. *Schriftenreihe für Vegetationskunde*, **25**, 3–185.

Lomolino, M.V. (1988) Body size of mammals on islands. *American Naturalist*, **125**, 310–6.

Long, G.E. (1977) Spatial dispersal in a biological control model for larch casebearer (*Coleophora laricella*). *Environmental Entomology*, **6**, 843–51.

Long, J.L. (1981) *Introduced Birds of the World*, David and Charles, London, UK.

Lonsdale, W.M. (1994). Inviting trouble: introduced pasture species in northern Australia. *Australian Journal of Ecology*, **19**, 345–54.

Lowe, V.P.W. (1993) The spread of the grey squirrel (*Sciurus carolinensis*) into Cumbria since 1960 and its present distribution. *Journal of Zoology*, **231**, 663–7.

Lubchenco, J (1986) Relative importance of competition and predation: early colonization by seaweeds in New England, in J. Diamond and T.J. Case (eds) *Community ecology*, Harper and Row, New York, USA, pp. 537–55.

Lubina, J.A. and Levin, S.A. (1988) Spread of a reinvading species: range expansion in the California Sea Otter. *American Naturalist*, **131**, 526–43.

Mabberley, D.J. (1987) *The Plant Book*, Cambridge University Press, Cambridge, UK.

Macan, T.T. and Worthington, E.B. (1951) *Life in Lakes and Rivers*, New Naturalist 15, Collins, London, UK.

MacArthur, R.H. and Wilson, E.O. (1967) *The Theory of Island Biogeography*, Princeton University Press, Princeton, NJ, USA.

Macdonald, D. (1991) Feral cat *Felis catus*, in G.B. Corbet and S. Harris (eds) *The Handbook of British Mammals*, 3rd edn, Blackwell Scientific Publications, Oxford, UK, pp. 437–40.

Macdonald, D. and Barrett, P. (1993) *Mammals of Britain & Europe*, Collins Field Guides, HarperCollins, London, UK.

MacDonald, G.M. (1993) Fossil pollen analysis and the reconstruction of plant invasion. *Advances in Ecological Research*, **24**, 67–110.

Macdonald, I.A.W., Kruger, F.J. and Ferrar, A.A. (eds) (1986) *The Ecology and Management of Biological Invasions in Southern Africa*, Oxford University Press, Cape Town, South Africa, pp. 209–25.

Macdonald, I.A.W., Loope, L.L., Usher, M.B. and Hamann, O. (1989) Wildlife conservation and the invasion of nature reserves by introduced species: a global perspective, in J.A. Drake, H.A. Mooney, F. di Castri, R.H. Groves, F.J. Kruger, M. Rejmánek and M. Williamson (eds) *Biological Invasions, a Global Perspective*, John Wiley & Sons, Chichester, UK, pp. 215–55.

Macdonald, I.A.W., Powrie, F.J. and Siegfried, W.R. (1986) The differential invasion of southern Africa's biomes and ecosystems by alien plants and animals, in I.A.W. Macdonald, F.J. Kruger, and A.A. Ferrar (eds) *The Ecology and Management of Biological Invasions in Southern Africa*, Oxford University Press, Cape Town, South Africa, pp. 209–25.

Mack, R.N. (1989) Temperate grasslands vulnerable to plant invasions: characteristics and consequences, in J.A. Drake, H.A. Mooney, F. di Castri, R.H. Groves, F.J. Kruger, M. Rejmánek and M. Williamson (eds) *Biological Invasions: a Global Perspective*, SCOPE 37, John Wiley & Sons, Chichester, UK, pp. 155–79.

Mack, R.N. Predicting the identity and fate of plant invaders: emergent and emerging approaches. *Biological Conservation* (in press).

Mahdi, A., Law, R. and Willis, A.J. (1989) Large niche overlaps among coexisting plant species in a limestone grassland. *Journal of Ecology*, **77**, 386–400.

Marchant, J.P., Hudson, R., Carter, S.P. and Whittington, P. (1990) *Population Trends in British Breeding Birds*, British Trust for Ornithology, Tring, UK.

Markin, G.P. and Yoshioka, E. (1992) Evaluating proposed biological control programs for introduced plants, in C.P. Stone, C.W. Smith and J.T. Tunison (eds) *Alien Plant Invasions in Native Ecosystems of Hawai'i: Management and Research*, University of Hawaii Cooperative National Park Resources Study Unit, Hononlulu, HA, USA, pp. 757–78.

Marks, T.C. and Mullins, P.H. (1990) The seed biology of *Spartina anglica*, in A.J. Gra and P.E.M. Benham (eds) Spartina anglica *– a research review*, ITE research publication no. 2, HMSO, London, UK, pp. 20–5.

Marshall, L.G., Webb, S.D., Sepkoski, J.J.Jr. and Raup, D.M. (1982) Mammalian evolution and the Great American Interchange. *Science*, **215**, 1351–7.

Marti, R. (1993) White-headed duck. *World Birdwatch*, **15**(3), 18–19.

Martin, P.S. and Klein, R.G. (eds) (1984) *Quaternary Extinctions*, University of Arizona Press, Tucson, AZ, USA.

Martinez, N.D. (1992) Constant connectance in community food webs. *American Naturalist*, **139**, 1208–18.

Martins, P.S. and Jain, S.K. (1979) The role of genetic variation in colonising ability of rose clover (*Trifolium hirtum* All.). *American Naturalist*, **114**, 591–5.

Matthews, L.H. (1952) *British Mammals*, New Naturalist 21, Collins, London, UK.

May, R.M. (1973) *Stability and Complexity in Model Ecosystems*, Princeton University Press, Princeton, NJ, USA.

Maynard Smith, J. (1983) The genetics of stasis and punctuation. *Annual Review of Genetics*, **17**, 11–25.

Mayr, E. (1963) *Animal Species and Evolution*, The Belknap Press of Harvard University Press, Cambridge, Massachusetts, USA.

Mayr, E. (1965) The nature of colonizations in birds, in H.G. Baker and G.L. Stebbins (eds) *The Genetics of Colonizing Species*, Academic Press, New York, NY, USA, pp. 29–47.

Mayr, E. and Short, L.L. (1970) *Species Taxa of North American Birds*, Publication the Nuttall Ornithological Club No. 9., MA, USA.

McCallum, H. and Dobson, A. (1995) Detecting disease and parasite threats to endangered species and ecosystems. *Trends in Ecology and Evolution*. **10**, 190–4.

McCormick, J.F. and Platt, R.B. (1980) Recovery of an Appalachian forest following the chestnut blight. *American Midland Naturalist*, **104**, 264–73.

McNaughton, S.J. and Wolf, L.L. (1970) Dominance and the niche in ecological systems. *Science*, **167**, 131–9.

Mee, L.D. (1992) The Black Sea in crisis: a need for concerted international action. *Ambio*, **21**, 278–86.

Middleton, A.D. (1931) *The Grey Squirrel*, Sidgwick & Jackson, London, UK.

Milberg, P. and Tyberg, T. (1993) Naïve birds and noble savages – a review of man-caused pre-historic extinctions of island birds. *Ecography*, **16**, 229–50.

Mills, E.L., Leach, J.H., Carlton, J.T. and Secor, C.L. (1993) Exotic species in the Great Lakes: a history of biotic crises and anthropogenic introductions. *Journal of Great Lakes Research*, **19**, 1–54.

Mitchell, A. (1982) *The Trees of Britain and Northern Europe*, Collins, London, England.

Mitchell, D.S. and Bowmer, K.H. (1990) Aquatic weed problems and management in Australia, in A.H. Pieterse and K.J. Murphy (eds) *Aquatic Weeds*, Oxford University Press, Oxford, UK, pp. 355–70.

Mitter, C. and Schneider, J.C. (1987) Genetic change and insect outbreaks, in P. Barbosa and J.C. Schultz (eds) *Insect Outbreaks*, Academic Press, San Diego, USA, pp. 505–28.

Moffatt, C.B. (1938) The mammals of Ireland. *Proceedings of the Royal Irish Academy B*, **218**, 61–128.

Mollison, D. (1977) Spatial contact models for ecological and epidemic spread (with discussion). *Journal of the Royal Statistical Society B*, **39**, 283–326.

Mollison, D. (1991) Dependence of epidemic and population velocities on basic parameters. *Mathematical Biosciences*, **107**, 255–87.

Montevecchi, W.A., Blundon, E., Coombes, G., Porter, J and Rice, P. (1978) Northern Fulmar breeding range extended to Baccalieu Island, Newfoundland. *The Canadian Field-naturalist*, **92**, 80–2.

Morrison, H.G. and Desrosiers, R.C. (1994) Siminan immunodeficiency viruses, in R.G. Webster and A. Granoff (eds) *Encyclopedia of Virology*, Academic Press, London, UK, pp. 1316–21.

Morton, R.D., Law, R., Pimm, S.L. and Drake, J.A. On models for assembling ecological communities. *Oikos* (in press).

Moulton, M.P. (1993) The all-or-none pattern in introduced Hawaiian passeriformes: the role of competition sustained. *American Naturalist*, **141**, 105–19.

Moulton, M.P. and Pimm, S.L. (1986) Species introductions to Hawaii, in H.A. Mooney and J.A. Drake (eds) *Ecology of Biological Invasions of North America and Hawaii,* Ecological Studies 58, Springer-Verlag, New York, New York, USA, pp. 231–49.

Munro, T. (1935) Note on musk-rats and other animals killed since the inception of the campaign against musk-rats in October 1932. *Scottish Naturalist*, **1935**, 11–16.

Murray, J.D. (1993) *Mathematical Biology*, 2nd edn, Biomathematics 19, Springer-Verlag, Berlin, Germany.

Murray, N.D. (1986) Rates of change in introduced organisms. *Proceedings of the Sixth International Symposium on the Biological Control of Weeds*, 191–9.

Mutlu, E., Bingel, F., Gücüc, A.C., Melnikov, V.V., Niermann, U., Ostr, N.A. and Zaika, V.E. (1994) Distribution of the new invader *Mnemiopsis* sp. and the resident

Aurelia aurita and *Pleurobrachia pileus* populations in the Black Sea in the years 1991–1993. *I.C.E.S. Journal of Marine Science*, **51**, 407–21.

Myers, J.H. (1984) Ecological compression of *Taxodium distichum* var. nutans by *Melaleuca quinquenervia* in southern Florida, in K.C.Ewell and H.T. Odum (eds), *Cypress Swamps*, University of Florida Press, Gainesville, FL, USA, pp. 358–64.

Myers, J.H. (1987) Population outbreaks of introduced insects: lessons from the biological control of weeds, in P. Barbosa and J.C. Schultz (eds) *Insect outbreaks*, Academic Press, San Diego, USA, pp. 173–93.

Myers, J.H. and Iyer, R. (1981) Phenotypic and genetic characteristics of the European cranefly following its introduction and spread in western North America. *Journal of Animal Ecology*, **50**, 519–32.

Myers, K., Parer, I., Wood, D. and Cooke, B.D. (1994) The rabbit in Australia, in H.V. Thompson and C.M King (eds) *The European Rabbit, the History and Biology of a Successful Colonizer*, Oxford University Press, Oxford, UK, pp. 108–157.

National Academy of Sciences (1987) *Introduction of Recombinant DNA-engineered Organisms into the Environment, Key Issues*, National Academy Press, Washington, DC, USA.

National Research Council (1989) *Field Testing Genetically Modified Organisms.* National Academy Press, Washington, DC, USA.

Nee, S., Read, A.F., Greenwood, J.J.D. and Harvey, P.H. (1991) The relationship between abundance and body size in British birds. *Nature*, **351**, 312–13.

Nettleship, D.N. and Montgomerie, R.D. (1974) The Northern Fulmar, *Fulmarus glacialis*, breeding in Newfoundland. *American Birds*, **28**, 16.

Newsome, A.E. and Noble, I.R. (1986) Ecological and physiological characters of invading species, in R.H. Groves and J.J. Burdon (eds) *Ecology of Biological Invasions: an Australian Perspective*, Australian Academy of Science, Canberra, Australia, pp. 1–20.

Niemelä, J.K. and Spence, J.R. (1991) Distribution and abundance of an exotic ground-beetle (Carabidae): a test of community impact. *Oikos*, **62**, 351–9.

Niemelä, J.K. and Spence, J.R. (1994) Distribution of forest dwelling carabids (Coleoptera): spatial scale and the concept of communities. *Ecography*, **17**, 166–75

O'Connor, T.P. (1992) Pets and pests in Roman and medieval Britain. *Mammal Review*, **22**, 107–13.

Odum, E.P. (1971) *Fundamentals of Ecology*, 3rd edn, Saunders, Philadelphia, PA, USA.

Oguto-Ohwayo, R. (1995) Diversity and stability of fish stocks in Lakes Victoria Kyoga and Nabugabo after establishment of introduced species, in T.J. Pitcher and P.J.B. Hart (eds) *The Impact of Species Changes in African Lakes*, Chapman & Hall London, UK, pp. 59–81.

Okubo, A. (1980) *Diffusion and Ecological Problems: Mathematical Models* Biomathematics 10, Springer-Verlag, Berlin, Germany.

Okubo, A. (1988) Diffusion-type models in avian range expansion. *Acta XIX Congressus Internationalis Ornithologici*, **1**, 1038–49.

Okubo, A., Maini, P.K., Williamson, M. and Murray, J.D. (1989) On the spatial spread of the grey squirrel in Britain. *Proceedings of the Royal Society B*, **238**, 113–25.

Oldroyd, B.P., Cornuet, J.-M., Rowe, D., Rinderer, T.E. and Crozier, R.H. (1995 Racial admixture of *Apis mellifera* in Tasmania, Australia: similarities and differences with natural hybrid zones in Europe. *Heredity*, **74**, 315–25.

Ollason, J.C. and Dunnet, G.M. (1988) Variation in breeding success in fulmars, i

T.H. Clutton-Brock (ed.) *Reproductive Success*, Chicago University Press, Chicago, USA, pp. 263–78.

Olson, S.L. and James, H.F. (1991) Descriptions of thirty-two new species of birds from the Hawaiian Islands: Part I. Non-passeriformes. *Ornithological Monographs of the American Ornithologists' Union*, **45**, 1–88.

Opie, I. and Opie, P. (1955) *The Oxford Nursery Rhyme Book*, Clarendon Press, Oxford, UK.

Orr, H.A. and Coyne, J.A. (1992) The genetics of adaptation: a reassessment. *American Naturalist*, **140**, 725–42.

Ostenfeld, C.H. (1908) On the immigration of *Biddulphia sinensis* Grev. and its occurrence in the North Sea during 1903–1907. *Meddelelser fra Kommissionen for Havundersøgelser, Serie: Plankton* **1**(6), 1–51.

Owen-Smith, N. (1987) Pleistocene extinctions: the pivotal role of megaherbivores. *Paleobiology*, **13**, 351–62.

Owen-Smith, R.N. (1988) *Megaherbivores: The Influence of Very Large Body Size on Ecology*, Cambridge University Press, Cambridge, UK.

Paine, R.T. (1992) Food-web analysis through field measurement of per capita interaction strength. *Nature*, **355**, 73–5.

Palese, P. and Garciá-Sastre (1994) Influenza viruses: molecular biology, in R.G. Webster and A. Granoff (eds) *Encyclopedia of Virology*, Academic Press, London, UK, pp. 715–22.

Palmer, T.K. (1973) The house finch and starling in relation to California's agriculture, in S.C. Kendeigh and J. Pinowski (eds) *Productivity, population dynamics and systematics of granivorous birds*, PWN – Polish Scientific Publishers, Warsaw, Poland, pp. 275–90.

Panetta, F.D. and Randall, R.P. (1994) An assessment of the colonizing ability of *Emex australis*. *Australian Journal of Ecology*, **19**, 76–82.

Pareira, M.S. (1980) The effects of shifts and drifts on the epidemiology of influenza in man. *Philosophical Transactions of the Royal Society B*, **288**, 423–32.

Parer, I., Conolly, D. and Sobey, W.R. (1981) Myxomatosis: the introduction of a highly virulent strain of myxomatosis into a wild rabbit population at Urana in New South Wales. *Australian Wildlife Research*, **8**, 613–26.

Parkin, D. and Cole, S.R. (1985) Genetic differentiation and rates of evolution in some introduced populations of the house sparrow, *Passer domesticus*, in Australia and New Zealand. *Journal of Heredity*, **54**, 15–23.

Patterson, K.D. (1986) *Pandemic Influenza 1700–1900, a Study in Historical Epidemiology*, Rowman & Littlefield, Totowa, NJ, USA.

Patterson, R.S. (1994) Biological control of introduced ant species, in D.F. Williams (ed.) *Exotic Ants, Biology Impact and Control of Introduced Species*, Westview Press, Boulder, CO, USA, pp. 293–307.

Patton, J.L. and Hafner, M.S. (1983) Biosystematics of the native rodents of the Galapagos archipelago, Ecuador, in R.I. Bowman, M. Berson and A.E. Leviton, (eds) *Patterns of evolution in Galapagos organisms*, Pacific Division of the American Association for the Advancement of Science, San Francisco, CA, USA, pp. 539–68.

Peet, R.K. (1992) Community structure and ecosystem function, in D.C. Glenn-Lewis, R.K. Peet and T.T. Veblen (eds) *Plant Succession: Theory and Prediction*, Chapman & Hall, London, UK, pp. 103–51.

Perring, F.H. (1968) *Critical Supplement to the Atlas of the British Flora*, Thomas Nelson and Sons, London, UK.

Perring, F.H. and Walters, S.M. (1962) *Atlas of the British Flora*, Thomas Nelson and Sons, London, UK.

Perrins, J., Fitter, A. and Williamson, M. (1993) Population biology and rates of invasion of three introduced *Impatiens* species in the British Isles. *Journal of Biogeography*, **20**, 33–44.

Perrins, J., Williamson, M. and Fitter, A. (1992a) A survey of differing views of weed classification: implications for regulation of introductions. *Biological Conservation*, **60**, 47–56.

Perrins, J., Williamson, M. and Fitter, A. (1992b) Do annual weeds have predictable characters? *Acta Oecologia*, **13**, 517–33

Peschken, D.P. (1972) *Chrysolina quadrigemina* (Coleoptera: Chrysomelidae) introduced from California to British Columbia against the weed *Hypericum perforatum*: a comparison of behaviour, physiology, and colour in association with post-colonization adaptation. *Canadian Entomologist*, **104**, 1689–98.

Peterson, R.L. (1966) *The Mammals of Eastern Canada*, Oxford University Press, Toronto, Canada.

Petren, K., Bolger, D.T. and Case, T.J. (1993) Mechanisms in the competitive success of an invading sexual gecko over an asexual native. *Science*, **259**, 354–7.

Phillips, D.M. (1995) Images in clinical medicine: *Enterococcus faecalis*. *New England Journal of Medicine*, **332**, 26.

Pieterse, A.H. and Murphy, K.J. (eds) (1990) *Aquatic weeds*, Oxford University Press, Oxford, UK.

Pilcher, J.R. (1993) Radiocarbon dating and the palynologist: a realistic approach to precision and accuracy, in F.M. Chambers (ed.) *Climate Change and Human Impact on the Landscape*, Chapman & Hall, London, UK, pp. 23–32.

Pimm, S.L (1991) *The Balance of Nature?*, University of Chicago Press, Chicago, IL, USA.

Pimm, S.L. (1982) *Food Webs*, Chapman & Hall, London, UK.

Pimm, S.L. (1987) The snake that ate Guam. *Trends in Ecology and Evolution*, **2**, 293–5.

Pimm, S.L. (1989) Theories of predicting success and impact of introduced species, in J.A. Drake, H.A. Mooney, F. di Castri, R.H. Groves, F.J. Kruger, M. Rejmánek and M. Williamson (eds) *Biological Invasions: a Global Perspective*, SCOPE 37, John Wiley & Sons, Chichester, UK pp. 351–67.

Pointier, J.P., Guyard, A. and Mosser, A. (1989) Biological control of *Biomphalaria glabrata* and *B. straminea* by the competitor snail *Thiara tuberculata* in a transmission site of schistosomiasis in Martinique, French West Indies. *Annals of Tropical Medicine and Parasitology*, **83**, 263–9.

Polis, G.A. (1991) Complex trophic interactions in deserts: an empirical critique of food web theory. *American Naturalist*, **138**, 123–55.

Polunin, O. and Stainton, A. (1984) *Flowers of the Himalaya*, Oxford University Press, Oxford, UK.

Por, F.D. (1978) *Lessepsian Migration: The Influx of Red Sea Biota into the Mediterranean by Way of the Suez Canal*, Ecological Studies 23, Springer-Verlag, Berlin, Germany.

Por, F.D. (1990) Lessepsian migration. An appraisal and new data. *Bulletin de l'Institut Océanographique (Monaco)*, **Special 7**, 1–10.

Post, W.M. and Pimm, S.L. (1983) Community assembly and food web stability. *Mathematical Biosciences*, **64**, 169–92.

Pratt, H.D., Bruner, P.L. and Berrett, D.G. (1987) *The Birds of Hawaii and the Tropical Pacific*, Princeton University Press, Princeton, NJ, USA.

Rabinowitz, D. (1981) Seven forms of rarity, in H. Synge (ed.) *The Biological Aspects of Rare Plant Conservation*, John Wiley & Sons, Chichester, England, pp. 205–17.

Rackham, O. (1986) *The History of the Countryside*, Dent, London, UK.

Ramakrishnan, P.S. (ed.) (1991) *Ecology of Biological Invasions in the Tropics*, International Scientific Publications, New Delhi, India.

Ramakrishnan, P.S. and Vitousek, P.M. (1989) Ecosystem-level processes and the consequences of biological invasions, in J.A. Drake, H.A. Mooney, F. di Castri, R.H. Groves, F.J. Kruger, M. Rejmánek and M. Williamson (eds) *Biological Invasions: a Global Perspective*, SCOPE 37, John Wiley & Sons, Chichester, UK pp. 281–300.

Raven, J. and Walters, M. (1956) *Mountain Flowers*, New Naturalist 33, Collins, London, UK.

Raybould, A.F. and Gray, A.J. (1993) Genetically modified crops and hybridization with wild relatives: a UK perspective. *Journal of Applied Ecology*, **30**, 199–219.

Raybould, A.F., Gray, A.J., Lawrence, M.J. and Marshall, D.F. (1991) The evolution of *Spartina anglica* C.E. Hubbard (Gramineae): origin and genetic variability. *Biological Journal of the Linnean Society*, **43**, 111–26.

Reimer, N.J. (1994) Distribution and impact of alien ants in vulnerable Hawaiian ecosytems, in D.F. Williams (ed.) *Exotic Ants, Biology Impact and Control of Introduced Species*, Westview Press, Boulder, CO, USA, pp. 11–22.

Rejmánek, M. (1989) Invasibility of plant communities, in J.A. Drake, H.A. Mooney, F. di Castri, R.H. Groves, F.J. Kruger, M. Rejmánek and M. Williamson (eds) *Biological Invasions, a Global Perspective*, John Wiley & Sons, Chichester, UK, pp. 369–88.

Rejmánek, M. (1995) What makes a species invasive?, in P. Pysek, K. Prach, M. Rejmánek and M. Wade (eds) *Plant Invasions – General aspects and Special Problems*, SPB Academic Publishing, Amsterdam, Netherlands, pp. 3–13.

Rejmánek, M. Species richness and resistance to invasions, in G.H. Orians, R. Dirzo and J.H. Cushman (eds) *Diversity and processes in tropical forest ecosytems*, Springer-Verlag, Berlin, Germany (in press).

Rejmánek, M. and Richardson, D.M. What attributes make some plant species more invasive? *Ecology* (in press).

Reynolds, J.C. (1985) Details of the geographic replacement of the red squirrel (*Sciurus vulgaris*) by the grey squirrel (*Sciurus carolinensis*) in eastern England. *Journal of Animal Ecology*, **54**, 149–62.

Reynolds, J.E., Gréboval, D.F. and Mannini, P. (1995) Thirty years on: the development of the Nile perch fishery in Lake Victoria, in T.J. Pitcher and P.J.B. Hart (eds) *The Impact of Species Changes in African Lakes*, Chapman & Hall, London, UK, pp. 181–214.

Richardson, B.J., Rogers, P.M. and Hewitt, G.M. (1980) Ecological genetics of the wild rabbit in Australia. II. Protein variation in British, French and Australian rabbits and geographical distribution of the variation in Australia. *Australian Journal of Biological Sciences*, **33**, 371–83.

Richardson, D.M., Macdonald, I.A.W., Holmes, P.M. and Cowling, R.M. (1992) Plant and animal invasions, in R. Cowling (ed.) *The Ecology of Fynbos*, Oxford University Press, Capetown, SA, pp. 271–308.

Richardson, D.M., Williams, P.A. and Hobbs, R.J. (1994) Pine invasions in the southern hemisphere: determinants of spread and invadability. *Journal of Biogeography*, **21**, 511–27.

Rinderer, T.E., Oldroyd, B.P. and Sheppard, W.S. (1993) Africanized bees in the U.S. *Scientific American*, December, 52–8.

Rinderer, T.E., Stelzer, J.A., Oldroyd, B.P., Buco, S.M. and Rubink, W.L. (1991) Hybridization between European and Africanized honey bees in the neotropical Yucatan peninsula. *Science*, **253**, 309–11.

Ritchie, J. (1920) *The Influence of Man on Animal Life in Scotland*, Cambridge University Press, Cambridge, UK.

Roberts, L.I.N. (1986) The practice of biological control – implications for conservation, science and the community. *The Weka*, **9**, 76–84.

Robinson, G.A. (1961) Continuous plankton records: contributions towards a plankton atlas of the north-eastern Atlantic and the North Sea. Part 1. Phytoplankton. *Bulletins of Marine Ecology*, **5**, 81–9.

Robinson, R. (1984) Cat, in I.L. Mason (ed.) *Evolution of Domesticated Animals*, Longman, London, UK, pp. 217–25.

Rodda, G.H., Fritts, T.H. and Conry, P.J. (1992) Origin and population growth of the brown tree snake, *Boiga irregularis*, on Guam. *Pacific Science*, **46**, 46–57.

Roderick, G.K. (1992) Postcolonization evolution of natural enemies, in W.C. Kauffman and J.E. Nechols (eds) *Selection Criteria and Ecological Consequences of Importing Natural Enemies*, Entomological Society of America, Lanham, MD, USA.

Rogers, P.M., Arthur, C.P. and Soriguer, R.C. (1994) The rabbit in continental Europe, in H.V. Thompson and C.M King (eds) *The European Rabbit, the History and Biology of a Successful Colonizer*, Oxford University Press, Oxford, UK, pp. 22–63.

Room, P. (1990) Ecology of a simple plant-herbivore system: biological control of *Salvinia*. *Trends in Ecology and Evolution*, **5**, 74–9.

Root, T. (1988) *Atlas of Wintering North American Birds*, University of Chicago Press, Chicago, IL, USA.

Ross, H.A. (1983) Genetic differentiation of starling (*Sturnus vulgaris*) populations in New Zealand and Great Britain. *Journal of Zoology (London)*, **201**, 351–62.

Roth, V.L. (1992) Inferences from allometry and fossils: dwarfing of elephants on islands. *Oxford Surveys in Evolutionary Biology*, **8**, 259–88.

Roughgarden, J. (1989) The structure and assembly of communities, in J. Roughgarden, R.M. May and S.A. Levin (eds) *Perspectives in Ecological Theory*, Princeton University Press, Princeton, NJ, USA.

Roy, J., Navas, M.L. and Sonié, L. (1991) Invasion by annual brome grasses: a case study challenging the homoclime approach to invasions, in R.H. Groves and F. di Castri (eds) *Biogeography of Mediterranean Invasions*, Cambridge University Press, Cambridge, UK, pp. 207–24.

Royama, T. (1992) *Analytical Population Dynamics*, Chapman & Hall, London, UK.

Russo, E. and Cove, D. (1995) *Genetic Engineering. Dreams and Nightmares*, Freeman, London, UK.

Sabath, M.D., Broughton, W.C. and Easteal, S. (1981) Expansion of the range of the introduced toad *Bufo marinus* in Australia from 1935 to 1974. *Copeia*, **3**, 676–80.

Safriel, U.N. and Ritte, U. (1986) Population biology of Suez Canal migration – which way, what kind of species and why, in S. Karlin and E. Nevo (eds) *Evolutionary Process and Theory*, Academic Press, New York, NY, USA, pp. 561–82.

Sailer, R.I. (1983) History of insect introductions, in C.L Wilson and C.L. Graham

(eds) *Exotic Plant Pests and North American Agriculture*, Academic Press, New York, USA, pp. 15–38.
Salisbury, E. (1942) *The Reproductive Capacity of Plants*, G. Bell and Sons, London, UK.
Salisbury, E. (1961) *Weeds and Aliens*, New Naturalist 43, Collins, London, UK.
Salomonsen, F. (1950) *Grønlands Fugle*. Munksgaard, København, Denmark.
Salomonsen, F. (1965) Geographic variation in the Fulmar (*Fulmarus glacialis*) and zones of the marine environment in the North Atlantic. *Auk*, **85**, 327–55.
Samways, M.J. (1988) Classical biological control and insect conservation: are they compatible? *Environmental Conservation*, **15**, 349–54.
Samways, M.J. (1994) *Insect Conservation Biology*, Chapman & Hall, London, UK.
Sandars, E. (1937) *A Beast Book for the Pocket*, Oxford University Press, London, UK.
Sands, D.P.A. and Schotz, M.(1985) Control or no control: a comparison of the feeding strategies of two *Salvinia* weevils, in E.S. Delfosse (ed.) *Proceedings of the VI International Symposium on the Biological Control of Weeds*, Agriculture Canada, Ottawa, Canada, pp. 19–25.
Sars, G.O. (1867) *Histoire Naturelle des Crustacés d'Eau Douce Norvège. 1. Malacostracés.* Bergen, Norway.
Savidge, J.A. (1987) Extinction of an island forest avifauna by an introduced snake. *Ecology*, **68**, 660–8.
Schäfer, W., Straney, D., Ciuffetti, L. Van Etten, H.D. and Yoder, O.C. (1989) One enzyme makes a fungal pathogen, but not a saprophyte, virulent on a new host plant. *Science*, **246**, 247–9.
Scheffer, M., Hosper, S.H., Meijeer, M.-L., Moss, B. and Jeppesen, E. (1993) Alternative equilibria in shallow lakes. *Trends in Ecology and Evolution*, **8**, 275–9.
Schoener, T.W. (1989) The ecological niche. *Symposia of the British Ecological Society*, **29**, 79–113.
Scott, J.K. and Panetta, F.D. (1993) Predicting the Australian weed status of southern African plants. *Journal of Biogeography*, **20**, 87–93.
Seidler, R.J. and Levin, M. (eds) (1994) Ecological implications of transgenic plant release. *Molecular Ecology*, **3**, 1–89
Sharpe, F.R. and Lotka, A.J. (1911) A problem in age-distribution. *Philosophical Magazine*, **21**, 435–8, reprinted in 1977 *Biomathematics*, **6**, 98–100.
Sharrock, J.T.R. (1976) *The Atlas of Breeding Birds in Britain and Ireland*, British Trust for Ornithology, Tring, UK.
Shaw, M.W. (1994) Modelling stochastic processes in plant ecology. *Annual Review of Phytopathology*, **32**, 523–44.
Shaw, M.W. (1995) Simulation of population expansion and spatial pattern when individual dispersal distributions do not decline exponentially with distance. *Proceedings of the Royal Society B*, **259**, 243–8.
Sheppard, W.S., Rinderer, T.E., Mazzoli, J.A., Stelzer, J.A. and Shimanuki, H. (1991) Gene flow between African- and European-derived honey bee populations in Argentina. *Nature*, **349**, 782–4.
Shorrocks, B. and Coates, D. (eds) (1993) *The Release of Genetically-Engineered Organisms*. British Ecological Society, Ecological Issues No. 4. Field Studies Council, Shrewsbury, UK.
Shorten, M. (1954) *Squirrels*, New Naturalist Monograph 12, Collins, London, UK.
Shushkina, E.A. and Musayeva, E.I. (1990) Structure of planktic community of the Black sea epipelagic zone and its variation caused by invasion of a new ctenophore species. *Oceanology*, **30**, 225–8 (in Russian, translation UDC 581.526.325).

Simberloff, D. (1981) Community effects of introduced species, in M. Nitecki (ed.) *Biotic Crises in Ecological and Evolutionary Time*, Academic Press, New York, USA, pp. 53–81.

Simberloff, D. (1986) Introduced insects: a biogeographic and systematic perspective, in H.A. Mooney and J.A. Drake (eds) *Ecology of Biological Invasions of North America and Hawaii*, Ecological Studies 58, Springer-Verlag, New York, USA, pp. 3–26.

Simberloff, D. (1989) Which insect introductions succeed and which fail?, in J.A. Drake, H.A. Mooney, F. di Castri, R.H. Groves, F.J. Kruger, M. Rejmánek and M. Williamson (eds) *Biological Invasions: a Global Perspective*, SCOPE 37, John Wiley & Sons, Chichester, UK pp. 61–75.

Simberloff, D. (1990) Community effects of biological introductions and their implications for restoration, in D.R. Towns, C.H. Daugherty and I.E.A. Atkinson (eds) *Ecological Restoration of New Zealand Islands*, Conservation publication no. 2, Department of Conservation, Wellington, NZ.

Simberloff, D. and Boecklen, W. (1991) Patterns of extinction in the introduced Hawaiian avifauna: a reexamination of the role of competition. *American Naturalist*, **138**, 300–27.

Simberloff, D. and Stilling, P. Risks of species introduced for biological control. *Biological Conservation* (in press).

Simpson, D.A. (1984) A short history of the introduction and spread of *Elodea* Michx. in the British Isles. *Watsonia*, **15**, 1–9.

Simpson, G.G. (1944) *Tempo and Mode in Evolution*, Columbia University Press, New York, NY, USA.

Skelcher, G. (1993) *North-west Squirrel Project, Final Report*, Cumbria Wildlife Trust, Ambleside, UK.

Skellam, J.G. (1951) Random dispersal in theoretical populations. *Biometrika*, **38**, 196–218.

Slatkin, M. (1985) Gene flow in natural populations. *Annual Review of Ecology and Systematics*, **16**, 393–430.

Smith, D.R. (1991) African bees in the Americas: insights from biogeography and genetics. *Trends in Ecology and Evolution*, **6**, 17–21.

Smith, S. (1995) The coypu in Britain. *British Wildlife*, **6**, 279–85.

Snow, D.W. (ed.) (1978) *An Atlas of Speciation in African Non-passerine Birds*, British Museum (Natural History), London, UK.

Spencer, C.N., McClelland, B.R. and Stanford, J.A. (1991) Shrimp stocking, salmon collapse and eagle displacement. *BioScience*, **41**, 14–21.

Spivak, M. (1991) The africanization process in Costa Rica, in M. Spivak, D.J.C. Fletcher and M.D. Breed (eds) *The 'African' Honey Bee*, Westview Press, Boulder, CO, USA, pp. 137–56.

Spivak, M. (1992) The relative success of Africanized and European honey-bees over a range of life zones in Costa Rica. *Journal of Applied Ecology*, **29**, 150–62.

Spivak, M., Fletcher, D.J.C. and Breed, M.D. (1991) Introduction, in M. Spivak, D.J.C. Fletcher and M.D. Breed (eds) *The 'African' Honey Bee*, Westview Press, Boulder, CO, USA, pp. 1–9.

St. Louis, V.L. and Barlow, J.C. (1988) Genetic differentiation among ancestral and introduced populations of the Eurasian tree sparrow (*Passer montanus*). *Evolution*, **42**, 266–76.

Stace, C. (1991) *New Flora of the British Isles*, Cambridge University Press, Cambridge, UK.

Steadman, D.W. (1995) Prehistoric extinctions of Pacific island birds: biodiversity meets zooarchaeology. *Science* **267**, 1123–31.
Stearns, S.C. (1992) *The Evolution of Life Histories*, Oxford University Press, Oxford, UK.
Stebbins, G.L. (1971) *Chromosomal Evolution in Higher Plants*, Edward Arnold, London, UK.
Stehli, F.G. and Webb, S.D. (eds) (1985) *The Great American Biotic Interchange*, Plenum, New York, NY, USA.
Stemberger, R.S. and Gilbert, J.J. (1985) Body size, food concentration, and population growth in planktonic rotifers. *Ecology*, **66**, 1151–9.
Sterba, G. (1962) *Freshwater Fishes of the World*, translated and revised by D.W. Tucker. Vista Books, London, UK.
Stodart, E. and Parer, I. (1988) *Colonisation of Australia by the Rabbit* Oryctolagus cuniculus *(L.)*, Project Report 6, CSIRO Division of Wildlife and Ecology, Lyneham, ACT, Australia.
Stone, C.P. (1989) Non-native land vertebrates, in C.P. Stone and D.B. Stone (eds) *Conservation Biology in Hawai'i*, University of Hawaii Press, Honolulu, HA, USA, pp. 88–95.
Stone, C.P. and Stone D.B. (eds) (1989) *Conservation Biology in Hawai'i*, University of Hawaii Press, Honolulu, HA, USA.
Stone, G.N. and Sunnucks, P. (1993) Genetic consequences of an invasion through a patchy environment – the cynipid gallwasp *Andricus quercuscalicis* (Hymenoptera: Cynipidae). *Molecular Ecology* **2**, 251–68.
Sukopp, H. and Sukopp, U. (1993) Ecological long-term effects of cultigens becoming feral and of naturalization of non-native species. *Experientia*, **49**, 210–18.
Summers-Smith, J.D. (1988) *The Sparrows*, Poyser, Calton, UK.
Sutherland, W.J. and Baillie, S.R. (1993) Patterns in the distribution, abundance and variation of bird populations. *Ibis*, **135**, 209–10.
Syrett, P., Hill, R.L. and Jessep, C.T. (1985) Conflict of interest in biological control of weeds in New Zealand, in E.S. Delfosse (ed.) *Proceedings of the VI International Symposium on the Biology of Weeds*, Agriculture Canada, Ottawa, Canada, pp. 391–7.
Tallis, J.H. (1991) *Plant Community History*, Chapman & Hall, London, UK.
Tattersall, W.M. and Tattersall, O.S. (1951) *The British Mysidacea*, The Ray Society, London, UK.
Tauber, C.A. and Tauber, M.J. (1987) Inheritance of seasonal cycles in *Chrysoperla* (Insecta: Neuroptera). *Genetical Research*, **49**, 215–23.
Taylor, D. (1992) Controlling weeds in natural areas in Hawai'i: a manager's perspective, in C.P. Stone, C.W. Smith and J.T. Tunison (eds) *Alien Plant Invasions in Native Ecosystems of Hawai'i: Management and Research*, University of Hawaii Cooperative National Park Resources Study Unit, Hononlulu, HA, USA, pp. 752–5.
Temple, S.A. (1977) Plant-animal mutualism: coevolution with dodo leads to near-extinction in plants. *Science*, **197**, 885–6.
Templeton, A.R. (1980) The theory of speciation *via* the founder principle. *Genetics*, **94**, 1011–38.
Tennant, L.E. (1994) The ecology of *Wasmannia auropunctata* in primary tropical rainforest in Costa Rica and Panama, in D.F. Williams (ed.) *Exotic Ants, Biology Impact and Control of Introduced Species*, Westview Press, Boulder, CO, USA, 80–90.
Theobald, D.V. (1926) The American grey squirrel (*Neosciurus carolinensis* Gmelin) in

Kent, Sussex and Surrey. *Bulletin of the South-eastern Agricultural College, Wye*, 4
3–20.
Thomas, L.K. Jr (1980) *The Impact of Three Exotic Plant Species on a Potomac island*
National Park Service Monograph series no. 13, US Department of the Interior
Washington, DC, USA.
Thompson, H.V. (1994) The rabbit in Britain, in H.V. Thompson and C.M King (eds
The European Rabbit, the History and Biology of a Successful Colonizer, Oxford
University Press, Oxford, UK, pp. 64–107.
Tiedje, J.M., Colwell, R.K., Grossman, Y.L., Hodson, R.E., Lenski, R.E., Mack, R.N
and Regal, P.J. (1989) The planned introduction of genetically engineered organisms: ecological considerations and recommendations. *Ecology*, **70**, 298–315.
Tilman, D. (1982) *Resource Competition and Community Structure*, Princeton University Press, Princeton, NJ, USA.
Timofeeff-Ressovsky, N.W., Vorontsov, N.N. and Yablakov, A.V. (1977) *An Outline of Evolutionary Theory*, Nauk, Moscow, Russia (in Russian).
Tittensor, A.M. (1977) Squirrels, in G.B. Corbet and H.N. Southern (eds) *The Handbook of British Mammals*, 2nd edn, Blackwell Scientific Publications, Oxford, UK
pp. 153–72.
Toft, G.O. (1983) Changes in the breeding seabird population in Rogaland, SW
Norway, during 1949–1979. *Cinclus*, **6**, 8–13.
Travis, J. (1993) Invader threatens Black, Azov Seas. *Science*, **262**, 1366–7.
Turner, C.E. (1985) Conflicting interests and biological control of weeds, in
E.S. Delfosse (ed.) *Proceedings of the VI International Symposium on the Biological Control of Weeds*, Agriculture Canada, Ottawa, Canada, pp. 203–25.
Turner, C.E., Pemberton, R.W. and Rosenthal, S.S. (1987) Host utilization of native
Cirsium thistles (Asteracceae) by the introduced weevil *Rhinocyllus conicus* (Coleoptera: Curculionidae) in California. *Environmental Entomology*, 16, 111–15.
Twigg, G.I. (1992) The black rat *Rattus rattus* in the United Kingdom in 1989. *Mammal Review*, **22**, 33–42.
U.S. Congress, Office of Technology Assessment (1993) *Harmful Non-indigenous Species in the United States*, OTA-F-565, U.S. Government Printing Office, Washington
DC, USA.
Ulbrich, J. (1930) *Die Bisamratte*, Heinrich, Dresden, Germany.
Ulloa-Chacon, P. and Cherix, D. (1994) Perspectives on control of little fire ant
(*Wasmannia auropunctata*) on the Galapagos islands, in D.F. Williams (ed.) *Exotic Ants, Biology Impact and Control of Introduced Species*, Westview Press, Boulder
CO, USA, pp. 63–72.
Usher, M.B. (1986) Invasibility and wildlife conservation: invasive species on nature
reserves. *Philosophical Transactions of the Royal Society B*, **314**, 695–710.
Usher, M.B. (1988) Biological invasions of nature reserves: a search for generalisations
Biological Conservation, **44**, 119–35.
Usher, M.B., Crawford, T.J. and Banwell, J.L. (1992) An American invasion of Great
Britain: the case of the native and alien (*Sciurus*) species. *Conservation Biology*, **6**
108–15.
Usher, M.B., Kruger, F.J., Macdonald, I.A.W., Loope, L.L. and Brockie, R.E. (1988
The ecology of biological invasions into nature reserves: an introduction. *Biological Conservation*, **44**, 1–8.
van den Bosch, F., Hengeveld, R. and Metz, J.A.J. (1992) Analysing the velocity of
animal range expansion. *Journal of Biogeography*, **19**, 133–50.

van den Bosch, F., Metz, J.A.J. and Diekman, O. (1990) The velocity of spatial population expansion. *Journal of Mathematical Biology*, **28**, 529–65.

van den Tweel, P.A. and Eijsackers, H. (1987) Black cherry, a pioneer species or 'forest pest'. *Proceeding of the Koninklijke Nederlandse Akademie van Wetenschappen*, **90**, 59–66.

van Riper, C. III, van Riper, S.G., Goff, M.L. and Laird, M. (1986) The epizootiology and ecological significance of malaria on the birds of Hawaii. *Ecological Monographs*, **56**, 327–44.

VanBlaricom, G.R. and Estes, J.A. (1988) *The Community Ecology of Sea Otters*, Ecological Studies 65, Springer-Verlag, New York, USA.

Vartanyan, S.L., Garutt, V.E. and Sher, A.V. (1993) Holocene dwarf mammoths from Wrangel Island in the Siberian Arctic. *Nature*, **362**, 337–40.

Venables, L.S.V. and Venables, U.M. (1955) *Birds and Mammals of Shetland*, Oliver and Boyd, Edinburgh, UK.

Verkaik, A.J. (1987) The muskrat in the Netherlands. *Proceedings of the Koninklijke Nederlands Akademie van Wetenschappen*, **90**, 67–72.

Vermeij, G.J. (1991a) When biotas meet: understanding biotic interchange. *Science*, **253**, 1099–104.

Vermeij, G.J. (1991b) Anatomy of an invasion: the trans-Arctic interchange. *Paleobiology*, **17**, 281–307.

Vinogradov, M.Y., Shushkina, E.A., Musayeva, E.I. and Sorokin, P.Y. (1989) A newly acclimated species in the Black Sea: the ctenophore *Mnemiopsis leidyi* (Ctenophora: Lobata). *Oceanology*, **29**, 220–4 (in Russian, translation UDC 591.524.12(26)).

Vitousek, P.M. (1986) Biological invasions and ecosystem properties: can species make a difference?, in H.A. Mooney and J.A. Drake (eds) *Ecology of Biological Invasions of North America and Hawaii*, Ecological Studies 58, Springer-Verlag, New York, USA, pp. 163–76.

Vitousek, P.M. (1990) Biological invasions and ecosystem processes: towards an integration of population biology and ecosystem studies. *Oikos*, **57**, 7–13.

Vitousek, P.M. and Walker, L.R. (1989) Biological invasion by *Myrica faya* in Hawai'i: plant demography, nitrogen fixation, ecosystem effects. *Ecological Monographs*, **59**, 247–65.

Vitousek, P.M., Walker, L.R., Whittaker, L.D., Mueller-Dombois, D. and Matson, P.A. (1987) Biological invasion by *Myrica faya* alters ecosystem development in Hawaii. *Science*, **238**, 802–4.

von Broembsen, S.L. (1989) Invasion of natural ecosystems by plant pathogens, in J.A. Drake, H.A. Mooney, F. di Castri, R.H. Groves, F.J. Kruger, M.Rejmánek and M. Williamson (eds) *Biological Invasions: a Global Perspective*, SCOPE 37, John Wiley & Sons, Chichester, UK pp. 77–83.

Voous, K.H. (1960) *Atlas of European Birds*, Nelson, London, UK.

Voous, K.H. (1973) List of recent holarctic bird species. Non-passerines. *Ibis*, **115**, 612–38.

Waage, J.K. and Greathead, D.J. (1988) Biological control: challenges and opportunities. *Philosophical Transactions of the Royal Society B*, **318**, 111–28.

Wallace, A.R. (1876) *The Geographical Distribution of Animals*. Macmillan and Co., London, UK.

Warheit, K.I. (1992) A review of the seabirds from the Tertiary of the North Pacific: plate tectonics, paleoceanography, and faunal change. *Paleobiology*, **18**, 401–24.

Warren, P.H. (1990) Variation in food-web structure: the determinants of connectance. *American Naturalist*, **136**, 689–700.
Warwick, T. (1934) The distribution of the muskrat (*Fiber zibethicus*) in the British Isles. *Journal of Animal Ecology*, **3**, 250–67.
Warwick, T. (1940) A contribution to the ecology of the musk-rat (*Ondatra zibethica*) in the British Isles. *Proceedings of the Zoological Society of London A*, **110**, 165–201.
Webb, D.A. (1966) Dispersal and establishment: what do we really know? in J.G. Hawkes (ed.), *Reproductive Biology and the Taxonomy of Vascular Plants*, BSBI conference report 9, Pergamon Press, Oxford, UK, pp. 93–102.
Webb, S.D. (1991) Ecogeography and the Great American Interchange. *Paleobiology*, **17**, 266–80.
Webster, R.G. (1994) Influenza viruses: general features, in R.G. Webster and A. Granoff *Encyclopedia of Virology*, Academic Press, London, UK.
Weeda, E.J. (1987) Invasions of vascular plants and mosses into the Netherlands. *Proceedings of the Koninklijke Nederlandse Akademie van Wetenschappen* C, **90**, 19–29.
Weiner, J. (1995) On the practice of ecology. *Journal of Ecology*, **83**, 153–8.
Weiss, P.W. and Milton, S. (1984) *Chrysanthemoides monilifera* and *Acacia longifolia* in Australia and South Africa. *Proceedings of the 4th International Conference on Mediterranean Ecosytems*, pp. 159–60.
Welcomme, R.L (1988) *International Introductions of Inland Aquatic Species* FAO Fisheries Technical Paper 294, FAO, Rome, Italy.
Welcomme, R.L. (1992) A history of international introductions of inland aquatic species. *International Council for the Exploration of the Sea Marine Science Symposia*, **194**, 3–14.
Whilde, A. (1993) *Threatened Mammals, Birds, Amphibians and Fish in Ireland, Irish Red Data Book 2: Vertebrates*, HMSO for the Office of Public Works (Dublin) and the Department of the Environment (Belfast), Belfast, UK.
Whiteaker, L.D. and Gardner, D.E. (1992) Firetree (*Myrica faya*) distribution in Hawai'i, in C.P. Stone, C.W. Smith and J.T. Tunison (eds) *Alien Plant Invasions in Native Ecosystems of Hawai'i: Management and Research*, University of Hawaii Cooperative National Park Resources Study Unit, Hononlulu, HA, USA, pp. 225–40.
Whittaker, R.H., Levin, S.A. and Root, R.B. (1973) Niche, habitat, and ecotope. *American Naturalist*, **107**, 321–38.
Williams, D.F. (ed.) (1994) *Exotic Ants, Biology Impact and Control of Introduced Species*, Westview Press, Boulder, CO, USA.
Williams, G.C. (1992) *Natural Selection*, Oxford University Press, New York, NY, USA.
Williams, K.S. (1986) Climatic influences on weeds and their herbivores: biological control of St John's Wort in British Columbia. *Proceedings of the VI International Symposium on the Biological Control of Weeds*, 127–32.
Williamson, M. (1957) An elementary theory of interspecific competition. *Nature*, **180**, 422–5.
Williamson, M. (1972) *The Analysis of Biological Populations*, Edward Arnold, London, UK.
Williamson, M. (1981) *Island Populations*, Oxford University Press, Oxford, UK.
Williamson, M. (1984) Sir Joseph Hooker's lecture on insular floras. *Biological Journal of the Linnean Society*, **22**, 55–77.

Williamson, M. (1987) Are communities ever stable? *Symposia of the British Ecological Society*, **26**, 353–71.

Williamson, M. (1988) Potential effects of recombinant DNA organisms on ecosystems and their components. *Trends in Ecology and Evolution*, **6**(4), S32–5.

Williamson, M. (1989a) Mathematical models of invasion, in J.A. Drake, H.A. Mooney, F. di Castri, R.H. Groves, F.J. Kruger, M. Rejmánek and M. Williamson (eds) *Biological Invasions: a Global Perspective*, SCOPE 37, John Wiley & Sons, Chichester, UK pp. 329–50.

Williamson, M. (1989b) Natural extinction on islands. *Philosophical Transactions of the Royal Society B*, **325**, 457–68.

Williamson, M. (1991) Biocontrol risks. *Nature*, **353**, 354.

Williamson, M. (1992) Environmental risks from the release of genetically modified organisms (GMOs) – the need for molecular ecology. *Molecular Ecology*, **1**, 3–8.

Williamson, M. (1993) Invaders, weeds and the risk from genetically modified organisms. *Experientia*, **49**, 219–24.

Williamson, M. (1994) Community response to transgenic plant release: predictions from British experience of invasive plants and feral crop plants. *Molecular Ecology*, **3**, 75–9.

Williamson, M. and Brown, K.C. (1986) The analysis and modelling of British invasions. *Philosophical Transactions of the Royal Society B*, **314**, 505–22.

Williamson, M. and Fitter, A. The varying success of invaders. *Ecology* (in press, a).

Williamson, M. and Fitter, A. The characters of succesful invaders. *Biological Conservation* (in press, b).

Wilson, E.O. (1992) *The Diversity of Life*, W.W. Norton & Company, New York, NY, USA.

Wilson, F. (1965) Biological control and the genetics of colonizing species, in H.G. Baker and G.L. Stebbins (eds) *The Genetics of Colonizing Species*, Academic Press, New York, USA, pp. 307–29.

Wilson, J.B., Hubbard, J.C.E. and Rapson, G.L. (1988) A comparison of the realised niche relations of species in New Zealand and Britain. *Oecologia*, **76**, 106–10.

Wilson, J.B., Rapson, G.L., Sykes, M.T., Watkins, A.J. and Williams, P.A. (1992) Distributions and climatic correlations of some exotic species along roadsides in South Island, New Zealand. *Journal of Biogeography*, **19**, 183–94.

Winemiller, K.O. (1989) Must connectance decline with species richness? *American Naturalist*, **134**, 960–8.

Winston, M.L. (1987) *The Biology of the Honey Bee*, Harvard University Press, Cambridge, MA, USA.

Winston, M.L. (1992) *Killer bees: The Africanized Honey Bee in the Americas*, Harvard University Press, Cambridge, MA, USA.

Witte, F., Goldschmidt, T. and Wanink, J.H. (1995) Dynamics of the haplochromine cichlid fauna and other ecological changes in the Mwanza Gulf of Lake Victoria, in T.J. Pitcher and P.J.B. Hart (eds) *The Impact of Species Changes in African Lakes*, Chapman & Hall, London, UK, pp. 83–110.

Witte, F., Goldschmidt, T., Goudswaard, P.C., Ligtvoet, W., van Oijen, M.J.P. and Wanik, J.H. (1992) Species extinction and concomitant ecological changes in Lake Victoria. *Netherlands Journal of Zoology*, **42**, 214–32.

Wynne-Edwards, V.C. (1952) Geographical variation in the bill of the Fulmar (*Fulmarus glacialis*). *The Scottish Naturalist*, **64**, 84–101.

Wynne-Edwards, V.C. (1962) *Animal Dispersal in Relation to Social Behaviour*, Oliver and Boyd, Edinburgh, UK.

Yodzis, P. (1988) The indeterminacy of ecological interactions as perceived through perturbation experiments. *Ecology*, **69**, 508–15.

Young, J.P.W. and Johnston, A.W.B. (1989) The evolution of specificity in the legume-rhizobium symbiosis. *Trends in Ecology and Evolution*, **4**, 341–9.

Zaitsev, Y.P. (1992) Biological aspects of western Black Sea coastal waters. *Journal of Fisheries Oceanography*, **2**, 180–9.

Zaret, T.M. and Paine, R.T. (1973) Species introduction in a tropical lake. *Science*, **182**, 449–55.

Author index

Subject index

Taxa indexed as genera or families are not cross-indexed under higher categories. Thus birds' only relates to collections of birds.